Real-Time Road Profile Identification and Monitoring

Theory and Application

Synthesis Lectures on Advances in Automotive Technology

Editor

Amir Khajepour, *University of Waterloo, Canada*

The automotive industry has entered a transformational period that will see an unprecedented evolution in the technological capabilities of vehicles. Significant advances in new manufacturing techniques, low-cost sensors, high processing power, and ubiquitous real-time access to information mean that vehicles are rapidly changing and growing in complexity. These new technologies—including the inevitable evolution toward autonomous vehicles—will ultimately deliver substantial benefits to drivers, passengers, and the environment. Synthesis Lectures on Advances in Automotive Technology Series is intended to introduce such new transformational technologies in the automotive industry to its readers.

Real-Time Road Profile Identification and Monitoring: Theory and Application
Yechen Qin, Hong Wang, Yanjun Huang, and Xiaolin Tang
2019

Noise and Torsional Vibration Analysis of Hybrid Vehicles
Xiaolin Tang, Yanjun Huang, Hong Wang, and Yechen Qin
2018

Smart Charging and Anti-Idling Systems
Yanjun Huang, Soheil Mohagheghi Fard, Milad Khazraee, Hong Wang, and Amir Khajepour
2018

Design and Avanced Robust Chassis Dynamics Control for X-by-Wire Unmanned Ground Vehicle
Jun Ni, Jibin Hu, and Changle Xiang
2018

Electrification of Heavy-Duty Construction Vehicles
Hong Wang, Yanjun Huang, Amir Khajepour, and Chuan Hu
2017

Vehicle Suspension System Technology and Design
Avesta Goodarzi and Amir Khajepour
2017

Real-Time Road Profile Identification and Monitoring: Theory and Application
Yechen Qin, Hong Wang, Yanjun Huang, and Xiaolin Tang

ISBN: 978-3-031-00371-4   paperback
ISBN: 978-3-031-01499-4   ebook
ISBN: 978-3-031-00004-1   hardcover

DOI 10.1007/978-3-031-01499-4

A Publication in the Springer series
SYNTHESIS LECTURES ON ADVANCES IN AUTOMOTIVE TECHNOLOGY

Lecture #5
Series Editor: Amir Khajepour, *University of Waterloo, Canada*
Series ISSN
Print 2576-8107   Electronic 2576-8131

Real-Time Road Profile Identification and Monitoring

Theory and Application

Yechen Qin
Beijing Institute of Technology, China

Hong Wang
University of Waterloo, Canada

Yanjun Huang
University of Waterloo, Canada

Xiaolin Tang
Chongqing University, China

SYNTHESIS LECTURES ON ADVANCES IN AUTOMOTIVE TECHNOLOGY
#5

ABSTRACT

Ever stringent vehicle safety legislation and consumer expectations inspire the improvement of vehicle dynamic performance, which result in a rising number of control strategies for vehicle dynamics that rely on driving conditions. Road profiles, as the primary excitation source of vehicle systems, play a critical role in vehicle dynamics and also in public transportation. Knowledge of precise road conditions can thus be of great assistance for vehicle companies and government departments to develop proper dynamic control algorithms, and to fix roads in a timely manner and at the minimum cost, respectively. As a result, developing easy-to-use and accurate road estimation methods are of great importance in terms of reducing the cost related to vehicles and road maintenance as well as improving passenger comfort and handling capacity. A few books have already been published on road profile modeling and the influence of road unevenness on vehicle response. However, there is still room to discuss road assessment methods based on vehicle response and how road conditions can be used to improve vehicle dynamics.

In this book, we use several generalized vehicle models to demonstrate the concepts, methods, and applications of vehicle response-based road estimation algorithms. In addition, necessary tools, algorithms, and methods are illustrated, and the benefits of the road estimation algorithms are evaluated. Furthermore, several case studies of controllable suspension systems to improve vehicle vertical dynamics are presented.

KEYWORDS

road estimation, road profile, road classification, time-frequency analysis, vehicle system responses, machine learning, controllable suspension system, semi-active control strategies

Contents

Acknowledgments

This book was supported by the National Natural Science Foundation of China (Grant No. 51805028) and China Postdoctoral Science Foundation (Grant No. 2016M600934). We are also thankful to Morgan & Claypool Publishers and the series editor for providing the opportunity for this book.

Yechen Qin, Hong Wang, Yanjun Huang, and Xiaolin Tang
January 2019

Nomenclature

AKF	Adaptive Kalman Filter
ANFIS	Adaptive Neuro Fuzzy Inference System
ASTO	Adaptive Super Twisting Observer
BPNN	Back Propagation Neural Network
COC	Clipped Optimal Control
FRC	Flexible Roller Contact
GA	Genetic Algorithm
HG	Highway with Gravel
HMPC	Hybrid Model Predictive Control
LC	Linear Complementarity
MI	Mutual Information
MIP	Mixed Integer Programming
MILP	Mixed Integer Linear Programming
MIQP	Mixed Integer Quadratic Programming
ML	Machine Learning
MLD	Mixed Logical Dynamical
MOOP	Multi-Objective Optimization
MPC	Model Predictive Control
MR	Magnetorheological
mRMR	Minimum Redundancy Maximum Relevance
NSGA-II	Non-dominated Sorting Genetic Algorithm-II
PA	Pasture
PC	Predictive Control

PNN	Probabilistic Neural Network
PSD	Power Spectral Density
PSO	Particle Swarm Optimization
PWA	Piecewise Affine
QP	Quadratic Program
RF	Random Forest
RMS	Root Mean Square
RS	Rattle Space
SH	Smooth Highway
SGA	Steepest Gradient Algorithm
SMA	Sprung Mass Acceleration
SNR	Signal to Noise Ratio
SR	Smooth Runway
SRA	Square Root of Amplitude
STD	Standard Derivation
TD	Tire Deflection
WPD	Wavelet Packet Decomposition

CHAPTER 1

Introduction

1.1 MOTIVATION

Driven by increasingly stringent vehicle safety legislation and consumer expectations, more and more advanced vehicle control strategies that rely on driving conditions have been proposed to enhance vehicle dynamic performance [1–10]. Road profile, also known as road roughness or road unevenness, is defined as the irregularities in the pavement surface.[1] Road profile is one of the primary inputs of vehicle systems, and can dramatically influence vehicle performance regarding ride comfort and road handling [11, 12]. Bad road conditions increase vehicle operating costs as well as transportation costs. Additionally, the increased loads from each axle can adversely impact the durability of roads.

In America, approximately 25% of the major roads in cities are substandard, leading to colossal vehicle maintenance costs of approximately 400 dollars per driver per year on average [13]. A 2013 report showed that road maintenance expenses were about 20 billion Euros in the EU each year [14]. In Germany, the annual road maintenance expenses will increase from 2.7 billion Euros in 2015 to 3.6 billion Euros in 2025 [15]. In order to investigate pavement structures and monitor road conditions, the U.S. Federal Highway Administration (FHA) started the long-term pavement performance (LTPP) program in 1987 [16]. LTPP defined the standard data collection procedures, with which experiments were categorized into two sets, namely, the general pavement studies (GPS) and the specific pavement studies (SPS). The GPS was designed for existing pavement and included nine road levels, and the SPS included ten levels, which were built for specific designs.

The importance of road profile information motivates new research and development of vehicle-road coupled systems [17]. Li et al. [18] proposed a route planning algorithm based on passengers' feeling, which cannot only reduce travel time but also improve overall ride comfort. In the proposed framework, both road profile and road anomaly information are used in a multi-objective optimization. Zhang et al. showed the effectiveness of using road estimation for connected vehicles [19, 20]. Several recently proposed algorithms can combine customer behavior and road profile conditions to acquire the customer loading, which can then be used for virtual test [21, 22].

This trend has resulted in increased consumer claims of improved passenger comfort and handling capacity. Nevertheless, better passenger comfort cannot be obtained without sacrificing road handling capabilities for passive suspension systems, which are widely used [23].

[1]We will use the term road profile in this book.

Controllable suspension systems, including active and semi-active systems, thus emerged in the 1960s to remedy this issue. Compared to active suspension systems, semi-active systems can achieve a similar ride comfort and have lower costs which make them more applicable to mass production vehicles [24–26]. Current semi-active control papers typically assume the driving conditions remain unchanged, which is far from reality. The reason for this is that superior ride comfort and handling capacity cannot be simultaneously satisfied even for semi-active suspension systems [27], and road conditions are extremely diverse in the real world. Road adaptive technology is viewed as a possible solution to this problem. The gain of these systems can be tuned as proposed by Hong et al. [28]. Road condition estimation thus plays an essential role in the commercial applications of semi-active suspension systems.

It is clear that precise road information will help in the development of advanced vehicle control systems and can assist in timely road maintenance. Sustainable and efficient management of road networks is highly necessary.

1.2   ROAD ESTIMATION REVIEW

Road profile estimation has attracted much attention from both vehicle manufacturers and governments in recent decades [29]. Most road estimation algorithms that are currently adopted can be classified into three categories based on the type of sensors [30].

1. Direct measurement. This method uses a specially designed instrument to measure road irregularities by maintaining contact between the instrument and the road surface.

2. Non-contact measurement. This type uses non-contact sensors, e.g., laser, radar, and ultrasonic to scan and generate graphs containing road profile information.

3. System response-based estimation. This is an indirect method which requires sensors to be installed on the vehicle. With mounted accelerometers and displacement sensors, this category applies transfer functions or observer techniques to estimate road conditions.

The characteristics of these three categories are graphically presented in Figure 1.1. It can be seen from Figure 1.1 that all of the methods above have advantages and disadvantages. For the first two methods, the high accuracy of the estimation is the most attractive attribute. However, the high cost and sensitivity to either instrument dynamic characteristics or environmental effects can restrict their application. Conversely, the increasing number of sensors that have been adapted for dynamic vehicle control makes the third method more applicable to mass-production passenger vehicles. The main drawbacks of the third method are the lower estimation accuracy compared to the first two methods, and the requirements for either an accurate model or plenty of training data. Although the vigorous development of autonomous vehicles will make the second category a promising approach in the next decades, in-depth investigation of the second method is worthwhile, especially for middle- to low-end vehicles. A detailed literature review of this topic will be presented in the next section.

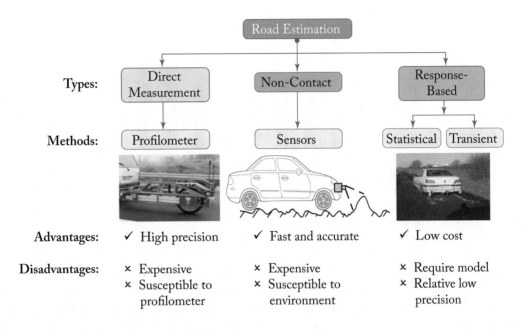

Figure 1.1: Road estimation categories [32, 40].

To illustrate the research trends in road profile estimation in both academia and industry, the number of the published papers in the databases of Web of knowledge (representing mostly academic research) and SAE (representing the focus of the automotive industry) are investigated and shown in Figure 1.2. It can be seen from Figure 1.2 that the number of published research papers in this area has continued to rise since 2001, which indicates the increasing research interest in this field.

1.2.1 DIRECT MEASUREMENTS

Direct measurement methods use equipment to measure the road profile by adhering to the ground. Several types of instruments have been applied to measure road profiles. Two representative examples of direct measurement equipment are the dipstick profiler and the profilograph [31].

A dipstick profiler has two tandem legs and two digital displays on which the relative elevation to another leg is provided. The operator will measure a pavement section by alternatingly pivoting the profiler about each leg, and the relative evaluation is recorded sequentially. The sample rate of such a profiler is usually approximately 0.25 Hz, and the resolution can reach 0.1 mm. Generally, this profiler is used to calibrate the results recorded by other sophisticated instruments.

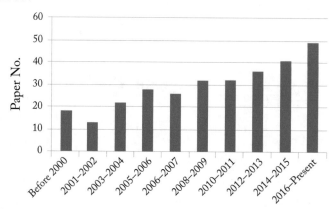

Figure 1.2: Research trend of road estimation.

A profilograph contains a mounted sensing wheel that can move vertically, and the road profile is recorded as per the motion of the wheel. The significant differences between different profilographs are the structure of the wheels and the data processing procedures. The minimum measureable spatial frequency for a typical profilograph is about 0.05 cycles/m. The profilograph was first proposed in the 1970s and is widely used for pavement measurement today. The French Road Research Laboratory developed the longitudinal profile analyzer, which had a measurement range of ±100 mm and a spatial frequency ranging from 0.05 cycles/m to 2 cycles/m [32]. Other famous profilographs include the TRRL (British Transport and Road Research Laboratory) profilometer, and the roughometer made by BPR (Bureau of Public Roads).

1.2.2   NON-CONTACT MEASUREMENTS

The recent vigorous development of autonomous vehicles has motivated the application of laser, light, and infrared transceivers, which can be used for road profile reconstruction in a non-contact manner. Generally, such road measurement systems contain non-contact sensors to measure the road surface and accelerometers are mounted in the body to account for the vertical movement of the vehicle [32]. For a system equipped with laser sensors, more than 5 lasers (up to 30 or even higher for different applications) distance meters are installed in the front of a vehicle. Each sensor can measure the distance between the light source to the road surface with a sample rate of up to 3500 Hz. Because of this, these systems can work at vehicle speeds in the range of 25–140 km/h. In addition to the laser sensors, these systems also include encoder sensors, accelerometers, and GPS to measure/calculate vehicle travel distance, vertical displacement, and geographic coordinates. Using these sensors, the road profile is obtained. Current commercialized laser displacement sensors for road profilometry can measure road profiles for up to 200 mm with an accuracy of 1% the measurement range [33]. Many papers have been published to describe this methodology [34, 35].

1.2.3 RESPONSE-BASED ESTIMATION

Although high accuracy can be expected with the first two methods, the reliance of these approaches on specific instruments restricts their practical application, especially in the middle- to low-end vehicles. Driven by increasingly stringent safety legislation and consumer expectations, more and more sensors have been used for system dynamic control. This facilitates the application of system response based road estimation. The generated road information from these systems, i.e., road level or road time domain profiles, can then be used for control parameter selection or adaptive variation of controller gains [36, 37]. A brief introduction to the development of response-based road estimation algorithms are presented below, with six representative publications from 1995–2015 being tabulated in Table 1.1.

Table 1.1: Representative publications for response based road estimation

Year	Authors	Keywords
1995	Hrovat et al.	Transfer function, rattle space [38]
1999	Yi et al.	Frequency estimation, sprung mass acceleration [24]
2002	Fialho et al.	Rattle space, threshold [39]
2006	Imine et al.	Sliding mode observer, time domain estimation [32]
2011	Doumiati et al.	Kalman filter, time domain estimation [40]
2015	Martinez et al.	Youla parameterization, time domain estimation [41]

In Table 1.1, we can see that research into response-based road estimation method began in 1995, when Hrovat et al. [38] used rattle space to probe road conditions. Then in 1999, Yi et al. [24] proposed a frequency domain-based estimation algorithm using the spring mass acceleration response. Fialho et al. [39] classified road conditions as rough or smooth by defining thresholds based on rattle space. Later in 2006, Imine et al. [32] used a sliding mode observer to estimate time domain road profiles and a comparison between the proposed algorithm and a profilograph was then carried out. Doumiati et al. [40] treated road excitation as an extended system state, and a Kalman filter was proposed to provide road profile estimation in real time. For semi-active suspension systems, Martinez et al. [41] used the Youla parameterization method, which was effective for variations in controller parameters. It can be seen from Table 1.1 that more and more complex road estimation algorithms have been proposed in recent decades, and methods that can provide accurate results are highly desired. Currently, the response-based road estimation algorithms can be divided into two categories, namely, road level classification and road profile estimation in the time domain. A more detailed overview of these two categories is given below.

Road classification algorithm. Road classification refers to the group of algorithms by which the levels or anomalies of road excitation are estimated.

Gonzalez et al. first proposed a transfer function based road classification algorithm with the sole measurement of sprung mass acceleration [42]. Wang et al. then extended this method by taking velocity variations into consideration, and an experimental validation was performed to validate the algorithm [43]. Gorges et al. used this method to estimate road Power Spectral Density (PSD), and then presented a novel frequency road classification algorithm by minimizing the square error between the estimated PSD and its corresponding road class [44]. Qin et al. recently presented a novel speed independent road classification algorithm, with which no prior information of the road excitation or training process was required. Both simulation and experimental results showed that the algorithm was robust to changes in system parameters and vehicle velocity [45]. Fauriat et al. [46] proposed a Kalman filter based road estimation method, which can probe road profile statistical characteristics, e.g., PSD. The important factors of influence for the algorithm were comprehensively analyzed.

The rapid development of machine learning (ML) algorithms has motivated their application in road classification. Ngwangwa et al. [47] provided a back propagation neural network (BPNN) based method to classify road faults and features at mining sites. The experimental results conducted on a haul truck showed good classification accuracy for different road faults. Gorges et al. used a binary decision tree to classify six types of road obstacles. Toad test results showed that more than 87% of anomalies were successfully detected with the proposed algorithm [48]. A deep neural network-based classifier was proposed by Qin et al. and simulation results indicated that unsprung mass acceleration (UMA) is the most suitable response to be used for road classification purposes [49]. Nitsche et al. [50] compared the performance of pavement roughness estimation for three different ML models. In that work, multilayer perceptron, support vector machine, and random forest models were all utilized using the sprung mass acceleration from the full vehicle model. Qin et al. first proposed time-frequency analysis-based road classification algorithms, and a feature reduction algorithm was used to improve classification accuracy [51].

The estimated road class or anomaly types can be used for advanced suspension control [28, 52], observation accuracy improvement [53–55], and controller parameters tuning [56].

Time domain road profile estimation algorithms.   Current response-based road profile estimation algorithms mainly use an observer to reconstruct road excitation. Commonly used observers include Kalman filters and unknown input observers.

Doumiati et al. proposed an extended-state Kalman filter, with which road profile could be estimated. Experimental validation was performed. The results were comparable in accuracy to those obtained using a profilograph. Kang et al. further used a discrete Kalman filter with unknown input to estimate road profiles in the time domain, and the results were compared to laser profilometer measurements obtained from a real vehicle test [57]. Wang et al. combined a model error criterion with a Kalman filter [58] and used simulation results to validate the proposed algorithm.

Qin et al. trained an inverse suspension system model with an adaptive neuro fuzzy inference system (ANFIS) and compared it with other modeling algorithms. The simulation results showed that the ANFIS inverse model performed the best [30]. One drawback of such a method is the requirement of a comprehensive training set, which can be difficult to obtain. Chaari et al. proposed an independent component analysis-based road profile estimation algorithm based on inverse dynamics of quarter-, half-, and full-car models [59].

The concept of higher order sliding mode (HOSM) has been widely used for controller and observer design [60, 61]. Rath et al. used an HOSM observer to provide an estimation of road profile and tire road friction [62]. Li et al. used a jump diffusion process estimator to detect road anomalies, which was validated by in-vehicle experiments [63]. One issue associated with observer-based road estimation algorithms is the unmeasurable system responses such as displacement and velocity of sprung and unsprung masses and tire deflection.

The estimated road profile can be used for vehicle preview control [64, 65] and route planning [18, 66].

1.3   SUMMARY

This chapter started with an introduction into the importance of road estimation for public transportation and dynamic performance enhancement. Then, a literature review of different road estimation methods was presented. A simple survey of both academic and industry research was used to illustrate the increasing interest in this field. Current road estimation algorithms include three distinct categories, namely contact measurement, non-contact measurement, and estimation based on system response. Compared to the other two categories, response-based road estimation methods are more applicable for mass production vehicles as they provide acceptable accuracy at a lower cost. Finally, the algorithms for road level and road profile estimation, and their application, were introduced.

CHAPTER 2

System Modeling

This chapter introduces the models used in this book, which includes road profile and vehicle system models.

2.1 ROAD PROFILE MODELING

The road profile can be theoretically defined as the distance between the road surface and the base surface, and its statistical characteristics are described using PSD [67]. To provide a comprehensive description of the road with a PSD, the Gaussian distribution assumption was first proposed by Dobbs and Robson in 1973 [68]. Subsequently, the ISO then modeled the road profile as a Gaussian stochastic process [67]. This assumption has been widely used in the literature that vehicle response is investigated under random road excitation [69–71]. Note that the actual road profile may not be accurately represented by a stationary Gaussian model. This is mainly caused by road anomalies such as speed bumps and potholes [72]. These events typically occur above $\pm 3\sigma$, and make the measured road data exhibit non-Gaussian properties [73]. Since the main purpose of this book is random road profile estimation instead of road anomaly detection, the Gaussian distribution assumption is adopted. The reader can refer to [29, 74–77] for further information on road anomaly detection and identification.

2.1.1 DEFINITION OF ROAD PROFILE WITH THE SAME PSD STRUCTURE

According to ISO 8601, the PSD of the random road profile can be defined as:

$$G_q(n) = G_q(n_0) \left(\frac{n}{n_0} \right)^{-W}, \tag{2.1}$$

where n is the spatial frequency, n_0 is the reference spatial frequency with the value of 0.1 m^{-1}, and $G_q(n)$ is the PSD value at spatial frequency n. The differences of various road levels are determined by the reference PSD $G_q(n_0)$, and larger $G_q(n_0)$ represents worse road conditions. W is the waviness, which reflects the frequency structure. A larger value of W indicates a larger amplitude in the low frequency range, and ISO 8608 defines $W = 2$. With both $G_q(n_0)$ and W, the statistical characteristics of the road can be uniquely determined. The definition of road level given by ISO 8608 is shown in Table 2.1.

Table 2.1: Road level definition given by ISO 8608

Level	$G_q(n_0)$ 10^{-6} m³	Level	$G_q(n_0)$ 10^{-6} m³
A	16	E	4,096
B	64	F	16,384
C	256	G	65,536
D	1,024	H	262,144

For a vehicle driving with a velocity of v, the spatial frequency $G_q(n)$ can be transformed into the time-frequency $G_q(f)$ according to Eq. (2.2):

$$G_q(f) = \frac{1}{v}G_q(n) = G_q(n_0)\, n_0^2 \frac{v}{f^2}, \qquad (2.2)$$

where $G_q(f)$ is the time-frequency PSD, and f is the time frequency in Hz. By defining the PSD coefficient as $R_r = G_q(n_0)\, n_0^2 v$, the relationship between vehicle velocity and excitation energy can be shown in Figure 2.1. Figure 2.1 reveals that increasing vehicle velocity will result in a larger R_r, and a proportional relationship can be observed.

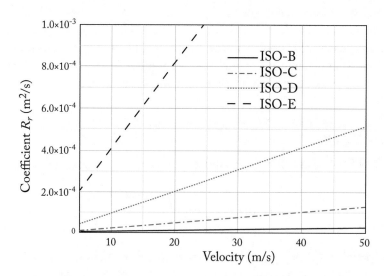

Figure 2.1: The relationship between vehicle velocity and excitation energy.

Note that the spatial frequency range is important to vehicle dynamic analysis, and this book defines the range to be $n \in [0.011, 2.83]$. When a vehicle is driving with a velocity between 36–108 km/h, this definition can cover the time frequency range of 0.33–28.3 Hz, which includes the sprung mass resonant frequency of 1–2 Hz and the unsprung mass resonant frequency of 10–15 Hz.

The generation of time domain road profile signals can be achieved using numerous methods including the rational function method [78], filtered white noise method [79], and integrated white noise method [80].

2.1.2   DEFINITION OF ROAD PROFILE WITH DIFFERENT PSD STRUCTURE

Although the ISO defined road assumes constant road waviness W, real-world roads may exhibit different frequency structures. Previous research has revealed that the W value of the pavements in the U.S., Germany, and Sweden varied from 1.6–2.4 with a mean value of 2 to 1.5–3.5, with a mean value of 2.5 in the past 30 years [81]. Similar to the PSD definition given in Eq. (2.1), the PSD of a road profile with a given waviness is described by the following equation [82]:

$$G_q(n) = C_{sp}n^{-W}, \tag{2.3}$$

where C_{sp} is the reference PSD at $n = 1$ m^{-1}. The parameters of six levels obtained from real-world road conditions with waviness varying from 1.5–3 are tabulated in Table 2.2. The abbreviations of different levels are given in parentheses.

Table 2.2: Definition of road with different PSD structure

Level	W	$C_{sp}10^{-6}$ m^3
Smooth Runway (SR)	3	4.3×10^{-5}
Smooth Highway (SH)	2.6	1.9×10^{-2}
ISO-B	2	6.4×10^{-1}
Highway with Gravel (HG)	2.1	4.4
ISO-E	2	41
Pasture (PA)	1.6	300

Figure 2.2 indicates that the larger W value will results in increased low-frequency components and lower excitation energy in the high-frequency components. Such variation in waviness presents unique challenges for road profile estimation. The following chapters will introduce various vehicle response based algorithms to estimate road profiles as defined in both Sections 2.1.1 and 2.1.2.

2.1.3   IRI AND ITS CORRELATION TO ISO INDEX

Many statistics for road quality evaluation have been proposed since the 1960s [68, 83]. One of the most well-received profile-based statistics is the international roughness index (IRI), which

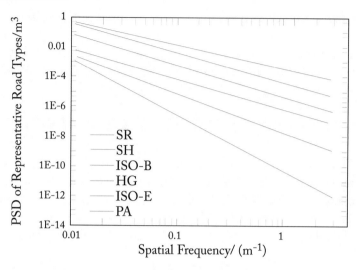

Figure 2.2: PSD of six road levels with non-identical waviness.

was developed by the World Bank in the 1980s and is widely used to evaluate and compare road roughness. This index uses the reaction of a single tire on a quarter vehicle model where the vehicle is traveling at a constant cruise speed of 50 mph (80 km/h). The IRI is defined as the average of the absolute value of rattle space over time. The detailed algorithm as well as model parameters can be found in reference [84].

Many researchers have investigated the correlations between the IRI and ISO indices. Kropac et al. provided an algorithm to transform the IRI to the ISO index and vice versa [85]. For a road profile defined by Eq. (2.3), the relationship between C_{sp} and IRI (mm/m) is given by:

$$IRI = a \sqrt{C_{sp}}, \tag{2.4}$$

where $a = 2.21$ for road waviness W equal to 2. For the range of waviness $W \in [1.5, 3]$, the coefficient a can be calculated as:

$$a = 2.21 \cdot \exp\left[-0.356\Delta W + 0.13(\Delta W)^2\right], \tag{2.5}$$

where $\Delta W = W - 2$. Alternatively, the reference PSD C_{sp} can be obtained according to Eq. (2.5) [86]:

$$C_{sp} = 0.205(IRI)^2 \cdot \exp\left[0.712\Delta W - 0.26\left(\Delta W\right)^2\right]. \tag{2.6}$$

We can observe that road waviness W is a crucial element of these transformations. Since the IRI relies on vehicle speed and waviness, Kropac formulated the following equation to represent such a relationship:

$$\log(IRI) = f\left(C_{sp}, W\right) + (0.5W - 1.5)\log(v), \tag{2.7}$$

where v is the vehicle velocity. For two different speeds where $v_1 \neq v_2$, the corresponding IRI indexes can be calculated as:

$$\log\left(IRI_1/IRI_2\right) = (0.5W - 1.5) \cdot \log\left(v_1/v_2\right). \tag{2.8}$$

The waviness can then be expressed as:

$$W = 3 + 2 \cdot \log\left(IRI_1/IRI_2\right) / \log\left(v_1/v_2\right). \tag{2.9}$$

Kropac then calculated the coefficients for $v_2 = 120,180$ km/h when $v_1 = 80$ km/h. Simulation results revealed the accuracy to be more than 95% [86].

2.1.4 ROAD PROFILE GENERATION

This section will discuss how to generate time domain road profile signals based on the PSD defined in Sections 2.1.1 and 2.1.2. Three widely used methods, namely harmonic function superposition [27], rational function [78], and integrated white noise [80], are subsequently introduced. Note that the first algorithm can be used for both road profile types given in Sections 2.1.1 and 2.1.2, and the last two are only applicable for roads with a constant PSD structure where waviness equals to 2.

1. Harmonic function superposition algorithm

   The basic idea of this algorithm is to approximate the road PSD with a series of harmonic functions. From Eq. (2.2), we can equally divide a frequency range $[f_1, f_2]$ into M parts. By replacing the $K^{th}(K = 1, 2, \ldots, M)$ frequency domain with its middle frequency f_{mid-K}, we can then approximate the PSD of Kth part as follows:

   $$p_K = G_q\left(f_{mid-K}\right) \cdot \Delta f_K \ (K = 1, 2, \ldots, M). \tag{2.10}$$

   According to the Plancherel theorem, the integral of a function's squared modulus is equal to the integral of the squared modulus of its frequency spectrum. Therefore, the amplitude of Kth part can be written as:

   $$A_K = \sqrt{p_K} = \sqrt{G_q\left(f_{mid-K}\right) \cdot \Delta f_K}. \tag{2.11}$$

   By summing M sine functions with frequency and amplitude defined by f_{mid-K} and Eq. (2.11), respectively, the generated road profile can be analytically expressed as:

   $$q(t) = \sum_{K=1}^{M} \sqrt{2 \cdot G_q\left(f_{mid-K}\right) \cdot \frac{f_2 - f_1}{M}} \sin\left(2\pi f_{mid-K}t + \Phi_K\right), \tag{2.12}$$

   where Φ_K is an independent and identically distributed random phase shift in the range $(0, 2\pi)$.

2. Rational function algorithm

Previous research has pointed out that the PSD described by Eq. (2.1) may result in over-estimation in the low-frequency range when compared with real-world data [23, 68]. Additionally, it has been shown that $G_q(n)$ will approach infinite when $n \to 0$. To remedy this, Michelberger et al. proposed a novel equation to represent the road PSD [78]:

$$G_q(n) = \frac{\alpha \rho^2}{\pi (\alpha^2 + n^2)},$$ (2.13)

where α and ρ are two constants. The values of α and ρ for different ISO levels are tabulated in Table 2.3 [87].

Table 2.3: α and ρ values for different ISO levels

Level	$\alpha/\mathrm{m^{-1}}$	ρ/mm	Level	$\alpha/\mathrm{m^{-1}}$	ρ/mm
A	0.111	37.4	E	0.111	603.2
B	0.111	75.4	F	0.111	1,206.4
C	0.111	150.8	G	0.111	2,412.8
D	0.111	301.6	H	0.111	4,825.6

The time domain road profile signal can then be generated as per the following equation:

$$\dot{q}(t) = -\alpha v q(t) + w(t),$$ (2.14)

where $w(t)$ is a white noise signal, and its covariance is calculated as:

$$\mathrm{cov}[w(t)] = E[w(t)w(t+\tau)] = 2\rho^2 \alpha v \delta(\tau),$$ (2.15)

where τ is time delay, and $\delta(\cdot)$ is a pulse function.

3. Integrated white noise algorithm

First, Eq. (2.2) is rewritten in the following form:

$$G_q(\omega) = 4\pi^2 G_q(n_0) \cdot n_0^2 \cdot \frac{v}{\omega^2}.$$ (2.16)

According to the spectral theorem, if a process $G(\omega)$ is stationary, then its PSD can be decomposed as follows:

$$G(\omega) = |H(j\omega)|^2,$$ (2.17)

$H(j\omega)$ can then be expressed as:

$$H(j\omega) = 2\pi n_0 \sqrt{G_{xr}(n_0)} v \frac{1}{j\omega}.$$ (2.18)

Consider the following equation:

$$\dot{q}(t) = \sigma \cdot w(t), \tag{2.19}$$

where $\sigma = 2\pi n_0 \sqrt{G_{x_r}(n_0)\, v}$ is the variance of the white noise signal. By comparing Eqs. (2.17) and (2.18), one can easily obtain:

$$q(t) = H_2(s)\, w(t), \quad H_2(s) = \frac{2\pi n_0 \sqrt{G_{x_r}(n_0)\, v}}{s}. \tag{2.20}$$

2.2 VEHICLE SYSTEM MODELING

This section starts by presenting a nonlinear quarter vehicle suspension model by considering its geometric characteristics. Then the equivalent linear quarter suspension model is derived using local linearization. Finally, the controllable damper model is presented.

2.2.1 NONLINEAR MACPHERSON SUSPENSION MODEL

Numerous suspension structures have been proposed and commercially applied. Some examples of these include the double wishbone, multi-link, swinging arm, and Macpherson models. Among these, the Macpherson-type suspension is the most widely adopted [88]. The reason for this is that this suspension structure is straightforward and can be preassembled into a unit. Further, it allows a larger space for engine layout by eliminating the upper control arm.

A schematic of a typical Macpherson suspension system is graphically represented in Figure 2.3 to illustrate the dynamics of a Macpherson strut [54]. A typical Macpherson strut is composed of a vehicle body, an axle, a tire, a coil spring, a damper, and a control arm. The vehicle sprung mass is assumed to have only a vertical motion. There are four degrees of freedom (DOF) in this system if the joint connecting the sprung mass and the suspension arm is assumed to be a bushing. The four DOFs include the sprung mass vertical displacement x_b, the suspension relative displacement z, the angular displacement of the lower wishbone θ_1 and the control arm θ_2, as shown in Figure 2.3a. If we further ignore the arm mass and assume the bushing to be a pin joint, then the new simplified nonlinear suspension model is shown in Figure 2.3b. In this case, there are only 2 DOFs, i.e., the sprung mass vertical movement and the control arm rotation angle [89].

For this model, we are given points A, B, and C with coordinates (y_A, x_A), (y_B, x_B), (y_C, x_C), respectively. The displacements of the sprung mass and road excitation amplitude are regarded as x_b and x_r, and the velocity and acceleration of x_b are defined as $\dot{x}_b$, $\ddot{x}_b$. Note that only the vertical movement of the sprung mass is considered. The angle α is the angle between the y axis and the axis OA. The angle θ is the angular displacement of the control arm with θ_0 being the initial value. The coordinates of the three points can be expressed as:

$$y_A = 0 \qquad x_A = x_b \tag{2.21}$$

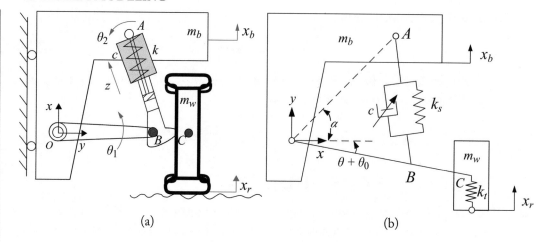

(a)   (b)

Figure 2.3: Macpherson-type suspension system: (a) schematic and (b) simplified system.

$$y_B = l_{OB} \left[\cos \left(\theta + \theta_0 \right) - \cos \left(\theta_0 \right) \right] \qquad x_B = x_b + l_{OB} \left[\sin \left(\theta + \theta_0 \right) - \sin \left(\theta_0 \right) \right] \qquad (2.22)$$

$$y_C = l_{OC} \left[\cos \left(\theta + \theta_0 \right) - \cos \left(\theta_0 \right) \right] \qquad x_C = x_b + l_{OC} \left[\sin \left(\theta + \theta_0 \right) - \sin \left(\theta_0 \right) \right]. \qquad (2.23)$$

Defining $\alpha' = \alpha + \theta_0$, the following relationships exist for triangle OAB.

$$l_{AB} = \sqrt{\left(l_{OA}^2 + l_{OB}^2 - 2 l_{OA} l_{OB} \cos \alpha' \right)} \qquad (2.24)$$

$$l'_{AB} = \sqrt{\left[l_{OA}^2 + l_{OB}^2 - 2 l_{OA} l_{OB} \cos \left(\alpha' - \theta \right) \right]}, \qquad (2.25)$$

where l_{AB} is the initial distance between points A and B, and l'_{AB} is the relative distance when α changes. By defining $\Delta l_{AB} = l_{AB} - l'_{AB}$, the derivative of Δl_{AB} is as follows:

$$\Delta \dot{l}_{AB} = \dot{l}_{AB} - \dot{l}'_{AB} = \frac{b_l \sin \left(\alpha' - \theta \right) \dot{\theta}}{2 \sqrt{a_l - b_l \cos \left(\alpha' - \theta \right)}}, \qquad (2.26)$$

where $a_l = l_{OA}^2 + l_{OB}^2$ and $b_l = 2 l_{OA} l_{OB}$. For the simplified system, the kinetic energy T, potential energy U and the Raleigh's dissipation function D can be calculated as:

$$T = \frac{1}{2} m_b \dot{x}_b^2 + \frac{1}{2} m_w \left(\dot{x}_{wx}^2 + \dot{x}_{wy}^2 \right) \qquad (2.27)$$

$$U = \frac{1}{2} k_s \left(\Delta l_{AB} \right)^2 + \frac{1}{2} k_t \left(x_w - x_r \right)^2 \qquad (2.28)$$

$$D = \frac{1}{2} c \Delta \dot{l}_{AB}^2, \qquad (2.29)$$

where m_b and m_w represent the sprung and the unsprung masses, respectively, and c is the damping coefficient. This coefficient can be either a constant, defined as c_p, for a linear damper or determined by any control law for a controllable damper (denoted as u). k_s and k_t are the suspension spring and tire stiffnesses, respectively.

By substituting Eqs. (2.21)–(2.26) into Eqs. (2.27)–(2.29), one can obtain the following equations:

$$T = \frac{1}{2}(m_b + m_w)\dot{x}_b^2 + \frac{1}{2}m_w l_{OC}^2 \dot{\theta}^2 + m_w l_{OC} \cos(\theta + \theta_0)\dot{\theta}\dot{x}_b \tag{2.30}$$

$$\begin{aligned} U = &\frac{1}{2}k_s\left\{2a_l - b_l\left[\cos\alpha' + \cos(\alpha' - \theta)\right]\right\} \\ &- k_s\sqrt{\left[a_l^2 - a_l b_l\left(\cos(\alpha' - \theta) + \cos\alpha'\right)\right] + b_l^2\cos\alpha'\cos(\alpha' - \theta)} \\ &+ \frac{1}{2}k_t\left\{x_b + l_{OC}\left[\sin(\theta + \theta_0) - \sin\theta_0\right] - x_r\right\}^2 \end{aligned} \tag{2.31}$$

$$D = \frac{cb_l^2\sin^2(\alpha' - \theta)\dot{\theta}^2}{8\left[a_l - b_l\cos(\alpha' - \theta)\right]}. \tag{2.32}$$

Based on the Lagrangian function, the dynamics equations can be derived from Eqs. (2.30)–(2.32) as follows:

$$\begin{aligned} &(m_b + m_w)\ddot{x}_b^2 + m_w l_{OC}\cos(\theta - \theta_0)\ddot{\theta} - m_w l_{OC}\sin(\theta - \theta_0)\dot{\theta}^2 \\ &+ k_t\left[x_b + l_{OC}\left(\sin(\theta - \theta_0) - \sin(-\theta_0) - x_r\right)\right] = 0, \end{aligned} \tag{2.33}$$

$$\begin{aligned} &m_w l_{OC}^2\ddot{\theta} + m_w l_{OC}\cos(\theta - \theta_0)\ddot{x}_b + k_t l_{OC}\cos(\theta - \theta_0) \\ &\quad\times\left\{x_b + l_{OC}\left[\sin(\theta - \theta_0) - \sin(-\theta_0)\right] - x_r\right\} \\ &\quad- \frac{1}{2}k_s\sin(\alpha' - \theta)\left[b_l + \frac{d_l}{\sqrt{c_l - d_l\cos(\alpha' - \theta)}}\right] = -l_{OB}c\Delta\dot{i}_{AB}, \end{aligned} \tag{2.34}$$

where $c_l = a_l^2 - a_l b_l\cos(\alpha + \theta_0)$, and $d_l = a_l b_l - b_l^2\cos(\alpha + \theta_0)$. Note that the approximations shown in Eqs. (2.35)–(2.36) are considered when using Eq. (2.34):

$$\frac{\partial\Delta i}{\partial\dot{\theta}} = \frac{b_l\sin(\alpha' - \theta)}{2\sqrt{a_l - b_l\cos(\alpha' - \theta)}} = \frac{l_{OA}l_{OB}\sin(\alpha' - \theta)}{l_{AB}} \approx l_{OB} \tag{2.35}$$

$$Q_j = -\frac{\partial D}{\partial\dot{\theta}} = -c\Delta\dot{i}\frac{\partial\Delta i}{\partial\dot{\theta}} \approx -l_{OB}c\Delta\dot{i}_{AB}. \tag{2.36}$$

Since the two variables $\ddot{x}_b$ and $\ddot{\theta}$ are coupled in Eqs. (2.33)–(2.34), the following method is used to decouple these equations [89]. First, define the system state variables as $x = [x_1\ x_2\ x_3\ x_4]^T = \begin{bmatrix} x_b & \dot{x}_b & \theta & \dot{\theta} \end{bmatrix}^T$, which are further defined as:

$$\begin{aligned} \dot{x}_1 &= x_2 & \dot{x}_2 &= f_1(x_1, x_2, x_3, x_4, c, x_r) \\ \dot{x}_3 &= x_4 & \dot{x}_4 &= f_2(x_1, x_2, x_3, x_4, c, x_r), \end{aligned}$$

where

$$f_1 = \frac{1}{D_1(x_3)} \left\{ m_w l_{OC}^2 \sin(x_3 - \theta_0) x_4^2 - \frac{1}{2} k_s \sin(\alpha' - x_3) \cos(x_3 - \theta_0) D_2(x_3) \right.$$
$$\left. - k_t l_{OC} \sin^2(x_3 - \theta_0) [x_1 + l_{OC}(\sin(x_3 - \theta_0) - \sin(-\theta_0)) - x_r] \right. \tag{2.37}$$
$$\left. + l_{OBC} \Delta i_{AB} \cos(x_3 - \theta_0) \right\}$$

$$f_2 = -\frac{1}{D_3(x_3)} \left\{ m_w^2 l_{OC}^2 \sin(x_3 - \theta_0) \cos(x_3 - \theta_0) x_4^2 - \frac{1}{2}(m_s + m_w) k_s \right.$$
$$\left. \times \sin(\alpha' - x_3) D_2(x_3) + m_s k_t l_{OC} \cos(x_3 - \theta_0) \right. \tag{2.38}$$
$$\left. \times \{x_1 + l_{OC}[\sin(x_3 - \theta_0) - \sin(-\theta_0)] - x_r\} + (m_s + m_w) l_{OBC} \Delta i_{AB} \right\},$$

and where

$$D_1(x_3) = m_b l_{OC} + m_w l_{OC} \sin^2(x_3 - \theta_0)$$
$$D_2(x_3) = b_l + \frac{d_l}{\sqrt{c_l - d_l \cos(\alpha' - x_3)}}$$
$$D_3(x_3) = m_b m_w l_{OC}^2 + m_w^2 l_{OC}^2 \sin^2(x_3 - \theta_0).$$

In this section, a nonlinear Macpherson type suspension system model was formulated. In the next section, this model will be linearized and a method to identify the equivalent parameters is presented.

2.2.2   SUSPENSION SYSTEM LINEARIZATION

A linear quarter model is frequently used for the design of suspension controllers and observers. The structure of the linear quarter model is depicted in Figure 2.4. We can see that the linear model contains the sprung mass, m_b, and the unsprung mass, m_w, with a spring k_s and a damper c_p positioned between them in parallel. The tire is simplified to be a linear spring k_t. It is of note that the reason why a linear tire model is considered is as follows.

1. The tire is a complicated system that possesses nonlinearity, time-delay, and uncertainties. These factors attract significant attention from both academia and the automobile industry, and many prominent publications have been published in recent decades [90–92]. Nevertheless, an accurate dynamic tire model that contains each of these factors is unavailable in most real-world applications. Therefore, a simpler tire model is more frequently used.

2. This book aims to present several accurate yet easy applied vehicle response-based road estimation methodologies. For the purpose of vertical suspension analysis, a tire is often

modeled as an amalgamation of spring and damper [93], and the damper is always uncared for [94, 95]. Later in Section 4.1.6, the influence of tire dynamics will be further discussed.

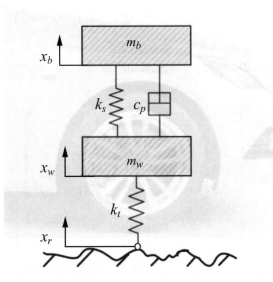

Figure 2.4: Structure of quarter vehicle model.

For a system with a linear damper, the dynamics equation can be formulated according to Newton's law of motion:

$$m_b\ddot{x}_b + k_s\,(x_b - x_w) + c_p\,(\dot{x}_b - \dot{x}_w) = 0$$
$$m_w\ddot{x}_w + k_t\,(x_w - x_r) - k_s\,(x_b - x_w) + c_p\,(\dot{x}_w - \dot{x}_b) = 0. \tag{2.39}$$

It can be seen that the linear model is equivalent to a particular case of the nonlinear model if we perform a Taylor expansion on the nonlinear model at the original point and set $l_{OB} = l_{OC}$ and $\theta_0 = 0$.

The equivalent parameters are then estimated based on the nonlinear model and experimental results obtained from a test rig. This book uses the least square estimation (LSE) method to estimate the parameters shown in Eq. (2.39). Equation (2.39) can be rewritten in the following form:

$$m_{beq}\ddot{x}_b + k_{seq}\,(x_b - x_w) + f_{deq} = \phi_1^T\vartheta \tag{2.40}$$
$$m_w\ddot{x}_w - k_{seq}\,(x_b - x_w) - f_{deq} + f_{tf} = \phi_2^T\vartheta, \tag{2.41}$$

where $f_{tf} = k_t\,(x_w - x_r)$ is the tire force, and k_{seq} and f_{deq} are the equivalent suspension spring stiffness and damper force, respectively. Since vehicle damping characteristics are always designed to be asymmetrical, i.e., bigger damping coefficient during rebound [88], f_{deq} is calculated by the following equation [96],

$$f_{deq} = 0.5c_{dleq} \left[1 + \text{sgn} \left(\dot{x}_b - \dot{x}_w \right) \right] \left(\dot{x}_b - \dot{x}_w \right)$$
$$+ 0.5c_{dyeq} \left[1 - \text{sgn} \left(\dot{x}_b - \dot{x}_w \right) \right] \left(\dot{x}_b - \dot{x}_w \right), \tag{2.42}$$

where c_{dleq} and c_{dyeq} are the equivalent coefficients for rebound and compression. ϕ_1, ϕ_2, and ϑ in Eqs. (2.40) and (2.41) are two measurement matrices and the vector of estimated parameters, which can be given as:

$$\phi_1 = \begin{bmatrix} \ddot{x}_b \\ 0 \\ x_b - x_w \\ 0.5 \left[1 + \text{sgn} \left(\dot{x}_b - \dot{x}_w \right) \right] \left(\dot{x}_b - \dot{x}_w \right) \\ 0.5 \left[1 - \text{sgn} \left(\dot{x}_b - \dot{x}_w \right) \right] \left(\dot{x}_b - \dot{x}_w \right) \end{bmatrix}$$

$$\phi_2 = \begin{bmatrix} 0 \\ \ddot{x}_w \\ - \left(x_b - x_w \right) \\ -0.5 \left[1 + \text{sgn} \left(\dot{x}_b - \dot{x}_w \right) \right] \left(\dot{x}_b - \dot{x}_w \right) \\ -0.5 \left[1 - \text{sgn} \left(\dot{x}_b - \dot{x}_w \right) \right] \left(\dot{x}_b - \dot{x}_w \right) \end{bmatrix}$$

$$\vartheta = \begin{bmatrix} m_{beq} \ m_{weq} \ k_{seq} \ c_{dleq} \ c_{dyeq} \end{bmatrix}^T.$$

Equations (2.40) and (2.41) can then be expressed as:

$$\phi_1^T \vartheta = 0 \tag{2.43}$$
$$\phi_2^T \vartheta - f_{tf} = 0. \tag{2.44}$$

The estimated parameters matrix is defined as $\hat{\vartheta} = \begin{bmatrix} \hat{m}_{beq} \ \hat{m}_{weq} \ \hat{k}_{seq} \ \hat{c}_{dleq} \ \hat{c}_{dyeq} \end{bmatrix}^T$, the estimation error matrix is then given by $\tilde{\vartheta} = \hat{\vartheta} - \vartheta$, and the estimation errors are then given by $\varepsilon_1 \equiv \phi_1^T \tilde{\vartheta}$, $\varepsilon_2 \equiv \phi_2^T \tilde{\vartheta} - f_t$. Assume the total sample number is N and define the error function E as follows:

$$E = \sum_{i=1}^{N} \left[\varepsilon_1^2(i) + \varepsilon_2^2(i) \right]. \tag{2.45}$$

The partial derivative of E w.r.t. $\hat{\vartheta}$ can be calculated as:

$$\frac{\partial E}{\partial \hat{\vartheta}} \Rightarrow \sum_{i=1}^{N} \left[\phi_1(i) \phi_1^T(i) + \phi_2(i) \phi_2^T(i) \right] \hat{\vartheta} = - \sum_{i=1}^{N} \left[\phi_2(i) f_{tf}(i) \right]. \tag{2.46}$$

By setting the partial derivative to zero, $\hat{\vartheta}$ can be written as

$$\hat{\vartheta} = -\left\{\sum_{i=1}^{N}\left[\phi_1(i)\phi_1^T(i) + \phi_2(i)\phi_2^T(i)\right]\right\}^{-1}\sum_{i=1}^{N}\left[\phi_2(i)f_{tf}(i)\right]. \tag{2.47}$$

With the LSE introduced in Eqs. (2.40)–(2.47), the unknown parameters $\hat{\vartheta}$ can be calculated with the data from a test rig, as shown in Figure 2.5.

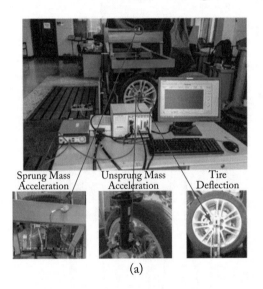

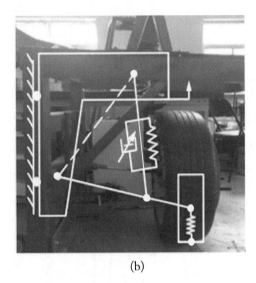

(a)   (b)

Figure 2.5: Graphical representation of quarter vehicle test rig: (a) instruments and (b) equivalence of nonlinear Macpherson model and test rig.

Figure 2.5a is a picture of the test rig. Two accelerometers are mounted on the sprung and unsprung masses to measure the accelerations, and an LVDT is placed between the tire axle and the excitation plate to perform tire deflection measurements. Additionally, an angular displacement sensor was installed on the control arm to estimate the rotation angle. The dynamic response of this system was measured and processed using a NI/Labview system with hardware including a PXI-6722 (Analog output) and a PXI-6224 (Analog input). A 2 Hz sine signal with 0.1 m amplitude was generated to estimate the parameters, and the estimated parameters are tabulated in Table 2.4.

With the parameters shown in Table 2.4, the measured and calculated responses for sprung mass acceleration and tire deflection, are compared below.

It can be seen from Figures 2.6 and 2.7 that the system responses calculated according to the estimated parameters better approximate the real nonlinear Macpherson-type suspension system than the model calculated using the nominal parameters.

Table 2.4: Estimated suspension parameters

Parameter	Nominal Value	Parameter	Estimated Value
m_b	350 kg	m_{beq}	324 kg
m_w	45 kg	m_{weq}	39 kg
k_s	22,000 N/m	k_{seq}	20,254 N/m
c_l	1,800 Ns/m	c_{dleq}	1,957 Ns/m
c_y	1,600 Ns/m	c_{dyeq}	1,677 Ns/m

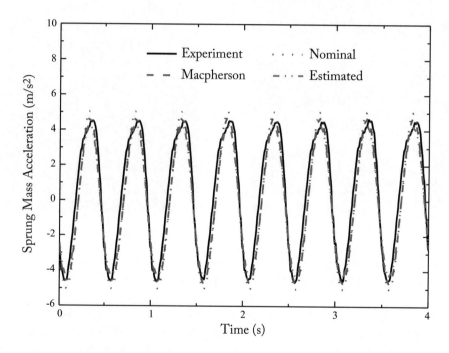

Figure 2.6: Comparison of results for sprung mass acceleration.

2.2.3   CONTROLLABLE DAMPER MODEL

In this section, a model for a controllable damper is presented. Many controllable dampers of different types have been commercialized in many mid-to high-end vehicles. A few examples are these are the Seville STS, the Acura MDX, and the GM-Lacrosse [97]. Currently, commercialized controllable dampers can be categorized into two types, namely magnetorheological (MR) dampers and proportional valve dampers. This book uses a proportional valve damper to generate the controllable damping force.

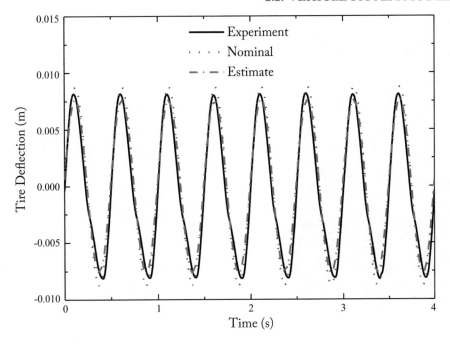

Figure 2.7: Comparison of results for tire deflection.

Generally, the damping force can be expressed by a function of the control command (which can be either a current or voltage—we use current in this book) and rattle space velocity. This is shown by

$$F_d = f\left(i, \dot{x}_b - \dot{x}_w\right). \tag{2.48}$$

In order to obtain the control current according to the estimated $\dot{x}_b - \dot{x}_w$ and the calculated control force, the inverse function is frequently used, which is described by

$$i = f^{-1}\left(F_d, \dot{x}_b - \dot{x}_w\right). \tag{2.49}$$

Many analytical and parametric methods have been proposed to depict this relationship [98, 99]. A nonparametric model proposed by Song et al. [100] is used in this book with the parameters being identified using experimental data. The characteristics of the investigated damper were tested with a MTS load frame, as shown in Figure 2.8a. A load cell and an LVDT sensor were installed on the test rig, and the relative velocity was calculated using a central difference method. Figure 2.8b shows the velocity-force map for the controllable damper.

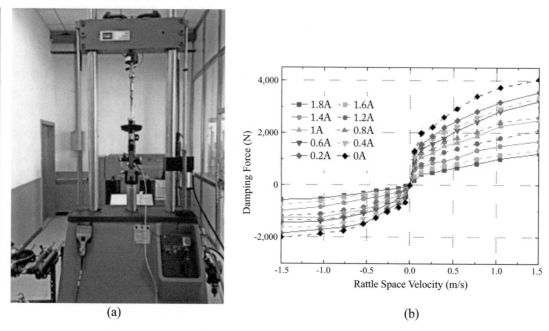

(a)   (b)

Figure 2.8: Controllable damper characteristics test: (a) damper mounted in MTS load frame and (b) velocity-force map.

Song et al. used the following equations to depict the relationship shown in Eq. (2.48):

$$F_d = A(i) \cdot S_b(\dot{x}_b - \dot{x}_w)$$

$$A(i) = \sum_{n=0}^{k} a_n i^n \qquad (2.50)$$

$$S_b(\dot{x}_b - \dot{x}_w) = \tanh\left[(b_1 i + b_0)(\dot{x}_b - \dot{x}_w)\right],$$

where n is the polynomial function order, a_n is the coefficient of nth order, and b_n are shape coefficients. These parameters were then identified based on particle swarm optimization (PSO) according to the data shown in Figure 2.8b. Since the characteristics are asymmetric, two separate models were created for positive and negative relative velocities. The objective function is selected as:

$$J = \frac{1}{2} \sum_{m=1}^{N} \left[F_d(m) - F_{test}(m)\right]^2, \qquad (2.51)$$

where F_{test} is the experimental data, and $N = 90$ is the total number of data points. The identified parameters for Eq. (2.50) are shown in Table 2.5.

Table 2.5: Estimated controllable damper parameters

Positive Model, $\dot{x}_b - \dot{x}_w > 0$		Negative Model, $\dot{x}_b - \dot{x}_w < 0$	
Parameter	Value	Parameter	Value
a_0^+	4,002	a_0^-	-2,002
a_1^+	-1,567	a_1^-	801.6
b_0^+	3.41	b_0^-	9.48
V_0^+	1.31	V_0^-	3.38

The inverse model can then be derived from Eq. (2.50), and the control current is described by

$$i = \begin{cases} \dfrac{F_{CDC} - a_0^+ S_b^+}{S_b^+ a_1^+}, & \dot{x}_b - \dot{x}_w > 0 \\ \dfrac{F_{CDC} - a_0^- S_b^-}{S_b^- a_1^-}, & \dot{x}_b - \dot{x}_w < 0. \end{cases} \tag{2.52}$$

Note that the model shown in Eqs. (2.50) and (2.52) cannot restrict the damper force, and the abnormal control force calculated by the controller may result in high control current, which is harmful to the damper. To solve this, a boundary damper model is formulated to limit the output force. The boundary forces that correspond to control currents of both 0 A and 1.8 A are piecewise fitted into several straight lines, as shown in Figure 2.9. The parameters for the lines are tabulated in Table 2.6.

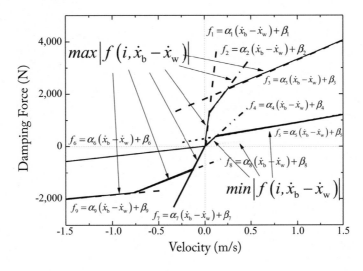

Figure 2.9: Boundary damper model.

Table 2.6: Parameters for the boundary damper model

$\dot{x}_b - \dot{x}_w$	f_1		f_2		f_3		f_4		f_5	
$\dot{x}_b - \dot{x}_w \geq 0$	α_1	β_1	α_2	β_2	α_3	β_3	α_4	β_4	α_5	β_5
	25,154	0	4,447	1,077	1,473	1,850	3,181	0	587	337
$\dot{x}_b - \dot{x}_w$	f_6		f_7		f_8		f_9			
$\dot{x}_b - \dot{x}_w \leq 0$	α_6	β_6	α_7	β_7	α_8	β_8	α_9	β_9		
	381	0	6,592	0	1,409	-674	351	-1,489		

2.3   SUMMARY

This chapter introduced the models that would be used in the subsequent chapters, which included road profile and vehicle system models. In the road profile model section, the correlation between ISO and IRI was first discussed. Then three different road generation algorithms were presented. For suspension system modeling, a nonlinear Macpherson suspension model was given and its linearization and the equivalent model were subsequently discussed. Finally, the controllable damper model was introduced which included inverse and boundary models of a commercially available controllable damper.

CHAPTER 3

Data-Driven Road Classification Algorithms

This chapter introduces data-driven road classification algorithms. The basic idea of the algorithm discussed in this chapter is to classify standard ISO road levels using system response levels of different roads. This chapter presents a feature-based road level classification algorithm. The features used are the time and frequency features from which a road level can be classified with the highest accuracy. In this part, we first present the differences in system responses on different road levels, and then illustrate how to select the most descriptive features from numerous candidate features. A simulation study is carried out to validate the proposed algorithm.

3.1 THE DIFFERENCE OF SYSTEM RESPONSES ON VARIOUS ROAD LEVELS

All the methods proposed in this chapter depend on differences in system response while a vehicle is driving on different road levels. Therefore, it is of great importance to investigate such differences at the very beginning. The sprung mass acceleration on different ISO road levels is shown in Figure 3.1. Note that all other responses, e.g., unsprung mass acceleration and rattle space, have similar characteristics to the sprung mass acceleration.

Figure 3.1 shows that the differences are obvious in both the time and frequency domains. Better classification accuracy can be expected if we can make full use of the information from both domains. Note that the road excitation is assumed to be a stationary random signal in Chapter 2. Therefore, we need to use statistical features to quantitatively describe the characteristics. Currently, numerous statistical features are used for this purpose, and they can be divided into two categories, namely, dimensional features and dimensionless features. Compared to the dimensional features, the dimensionless features can reflect signal distribution and are not affected by the mean value. Table 3.1 lists 11 features that can be used for signal description [101].

In Table 3.1, N is total number of data points, features f_1 and $f_3 - f_5$ represent signal energy, f_2, f_6, and f_7 are features describing signal distribution, f_8, f_9, and f_{11} indicate the influence of impact and impulse input.

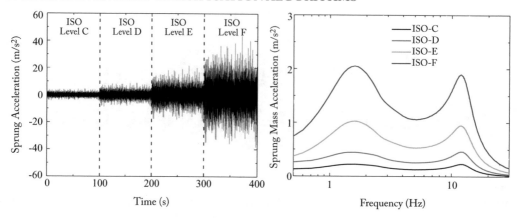

Figure 3.1: Difference of system sprung mass acceleration on various road levels: (a) time domain and (b) frequency domain.

3.2   FEATURES DEFINITION

In this section, we first define the candidate features in both the time and frequency domains. Next, the details of the selection of the most descriptive features are given.

All of the features shown in Table 3.1 can be used to describe signal characteristics, however, not all can effectively show the difference in response due to various road levels. In the authors' previous article [102], an improved distance evaluation method used by Lei et al. [101] was applied to select the most suitable features for road classification. The simulation results indicated that variance, square root of amplitude (SRA), maximum, and root mean square (RMS) are the most suitable features to describe differences in road profile. These are defined below:

$$\text{Variance} = \sqrt{\frac{\sum\limits_{n=1}^{N} (x(n) - \bar{x})^2}{N - 1}}, \quad \text{SRA} = \left(\frac{\sum\limits_{n=1}^{N} \sqrt{|x(n)|}}{N}\right)^2,$$

$$\text{RMS} = \sqrt{\frac{\sum\limits_{n=1}^{N} (x(n))^2}{N}}, \quad \text{Max} = \max |x(n)|.$$

It can be seen that all of these four features are dimensional parameters, which indicates that dimensional features are more suitable to assess road profiles from statistical signals compared to dimensionless features.

According to the definition of road PSD described by Eq. (2.1), the road profile has a higher amplitude in the low frequency range, and a lower amplitude in the high frequency range.

Table 3.1: Eleven representative features

No.	Feature Name	Expression	No	Feature Name	Expression
f_1	Mean value	$\dfrac{\sum\limits_{n=1}^{N} x(n)}{N}$	f_7	Kurtosis	$\dfrac{\sum\limits_{n=1}^{N} (x(n)-f_1)^4}{(N-1)f_2^4}$
f_2	Variance	$\sqrt{\dfrac{\sum\limits_{n=1}^{N} (x(n)-f_1)^2}{N-1}}$	f_8	Crest factor	$\dfrac{f_5}{f_4}$
f_3	Square root of amplitude	$\left(\dfrac{\sum\limits_{n=1}^{N} \sqrt{|x(n)|}}{N}\right)^2$	f_9	Clearance factor	$\dfrac{f_5}{f_3}$
f_4	Root mean square	$\sqrt{\dfrac{\sum\limits_{n=1}^{N} (|x(n)|)^2}{N}}$	f_{10}	Shape factor	$\dfrac{f_4}{\frac{1}{N}\sum\limits_{n=1}^{N}|x(n)|}$
f_5	Maximum	$\max |x(n)|$	f_{11}	Impulse factor	$\dfrac{f_5}{\frac{1}{N}\sum\limits_{n=1}^{N}|x(n)|}$
f_6	Skewness	$\dfrac{\sum\limits_{n=1}^{N} (x(n)-f_1)^3}{(N-1)f_2^3}$			

Therefore, wavelet packet decomposition (WPD) analysis can be used to decompose the signal into several frequency components. The structure of the adopted six-layer WPD is depicted in Figure 3.2. As mentioned in Section 2.1.4, the frequency range of interest for the suspension system is from 0.33–28.3 Hz. The lower- and upper-frequency bounds for the proposed six-layer WPD are set to be 0.8 Hz and 31.3 Hz. The blue boxes in Figure 3.2 indicate the frequency components of interest.

Taking the 4 candidate features and the 6 frequency ranges into consideration, a total of 29 features are generated and the index of all the features are tabulated in Table 3.2. The generated features are summarized as follows:

1. 4 features in the time domain;

2. 24 features in the frequency domain: the 4 candidate features for each of the 6 frequency range; and

3. 1 energy feature: the sum of squared terms for the 6 frequency components.

$$Energy = \sum_{i=2}^{6} \|aD_i\|^2 + \|dA_3\|^2 .$$

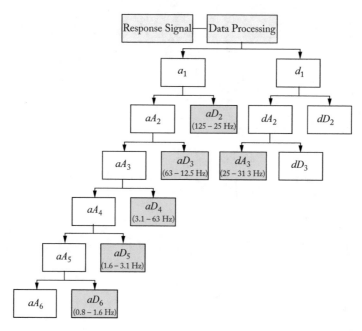

Figure 3.2: Six-layer WPD.

Table 3.2: Features definition

Features		Variance	SRA	RMS	Max	Energy
Time Domain		1	2	3	4	N/A
Frequency Domain	*Energy*	N/A	N/A	N/A	N/A	5
	aD2	6	12	18	24	N/A
	aD3	7	13	19	25	N/A
	aD4	8	14	20	26	N/A
	aD5	9	15	21	27	N/A
	aD6	10	16	22	28	N/A
	dA3	11	17	23	29	N/A

3.3 ROAD CLASSIFICATION ALGORITHM

3.3.1 OVERALL STRUCTURE

The overall structure of the proposed road classification algorithm is illustrated in Figure 3.3.

The proposed algorithm is composed of four stages, namely, signal pre-processing, feature combination, feature selection, and road classification. Section 3.2 defined how the features are calculated and the other stages will be discussed in this section.

3.3.2 SIGNAL PRE-PROCESSING

In this book, three measurable responses, i.e., SMA, RS, and UMA are used for classification analysis. In the signal pre-processing step, all the three responses are processed with the same procedure: low pass filtering, framing, and windowing.

Low-pass filtering. The sampling rate is set at 100 Hz with the upper frequency bound equal to 31.3 Hz. A low-pass filter is first applied to avoid signal aliasing, and the cut-off frequency is set at 50 Hz.

Framing. The classification interval in this part is set to 1 s. This means that 100 data points can be used for classification. Since the proposed algorithm is based on differences in the statistical features of various road levels, an overlap is required to ensure that enough information can be derived from the response signal. To accommodate this requirement, the length of the frame for any response signal is defined to be 200 points, including 100 points from the previous second.

Windowing. A Hamming window is used to prevent spectral leak at both the beginning and ending of each frame. The window is described by

$$Z(k) = 0.54 - 0.46\cos\left(2\pi\frac{k}{N-1}\right), \quad k = 0, 1, \ldots, N-1, \tag{3.1}$$

where $N = 200$ is the frame length, and the classification of the available signal is calculated by multiplying by $z(t)$.

3.3.3 FEATURE REDUCTION

We have defined the 29 candidate features in Table 3.2, and all of these candidate features can be used for road classification. Nevertheless, a dimensionality problem may arise if all candidate features are utilized. The reason for this can be stated as follows. First, the complexity of the trained classifier fully depends on the number of input variables. More input variables results in increased computation time and the curse of high-dimensionality. Second, traditional feature reduction methods mainly focus on the relative distance between different features and simple combination of features cannot ensure the optimality of the trained classifier. For these reasons, a feature reduction algorithm that can account for relevance and mutual redundancy is highly desired.

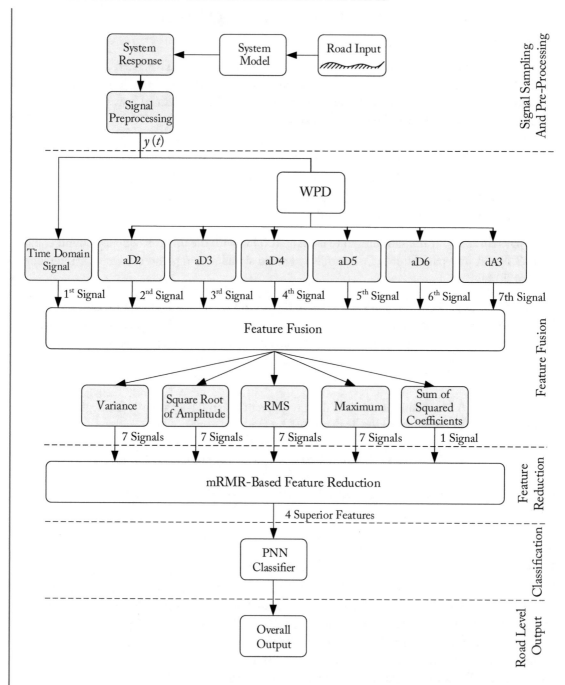

Figure 3.3: Overall structure for the superior feature based road classification algorithm.

In this section, a method called minimum redundancy maximum relevance (mRMR) proposed by Peng et al. is used to perform feature reduction [103, 104]. The mRMR approach defines the term mutual information (MI) to provide a quantitative description of two variables, which can be described by

$$I(m,n) = \sum_{i,j} p(m_i, n_j) \log \frac{p(m_i, n_j)}{p(m_i) p(n_j)}, \tag{3.2}$$

where $p(m,n)$, $p(m)$, and $p(n)$ are the joint probabilistic distributions for the marginal probabilities of variables m and n. For two variables with a length of M, the three probabilities can be estimated using the Gaussian kernel estimator. These calculations are shown in the following equations:

$$p(m,n) = \frac{1}{M} \sum_{i=1}^{M} \frac{1}{2\pi h^2} e^{-\frac{1}{2h^2}\left[(m-m_i)^2 + (n-n_i)^2\right]} \tag{3.3}$$

$$p(m) = \frac{1}{M} \sum_{i=1}^{M} \frac{1}{\sqrt{2\pi h^2}} e^{-\frac{1}{2h^2}(m-m_i)^2} \tag{3.4}$$

$$p(n) = \frac{1}{M} \sum_{i=1}^{M} \frac{1}{\sqrt{2\pi h^2}} e^{-\frac{1}{2h^2}(n-n_i)^2}, \tag{3.5}$$

where h is a constant for tuning the width of the kernels.

The minimum redundancy condition ensures that for a feature set S, all features, p, have minimal redundancy. This condition is described by

$$\min W_I, W_I = \frac{1}{|S|^2} \sum_{i,j \in S} I(g_i, g_j), \tag{3.6}$$

where $|S| = p$ is the feature number and g_i represents feature i.

The MI of $I(h, g_i)$ between the target class $h = \{h_1, h_2, \ldots, h_k\}$ and feature g_i is then calculated to maximize the relevance of all features in the set S. The maximum relevance condition can then be calculated using

$$\max V_I, V_I = \frac{1}{|S|^2} \sum_{i \in S} I(h, g_j). \tag{3.7}$$

The mRMR superior feature set can now be obtained by simultaneously optimizing both conditions defined in Eqs. (3.6) and (3.7). Finally, the following criterion is used to solve the contradiction between the two opposite conditions:

$$\max(V_I / W_I).$$

3.3.4   PNN CLASSIFIER

Specht et al. proposed Probabilistic Neural Network (PNN) in the 1990s [105]. Since then, it has been widely used to solve classification problems. PNN develops a distribution function and classifies the data based on the likelihood of features of different levels. The basic idea of PNN is to combine the well-known RBFNN and Bayesian theory. The Bayesian formula can be given in the following form:

$$P\left(X|Y\right) = \frac{P\left(Y|X\right)P\left(X\right)}{P\left(Y\right)},$$

where $P\left(X\right)$ and $P\left(Y\right)$ represent the known prior probabilities of X and Y. The posterior probability of $P\left(X|Y\right)$, which represents the probability of Y with existing X, should be known in advance. For the classification problem discussed in this chapter, Y represents the possibility that a pattern can be placed, and X is the pattern. The PNN algorithm then puts the pattern into a level with the highest probability. The structure of a PNN is illustrated in Figure 3.4.

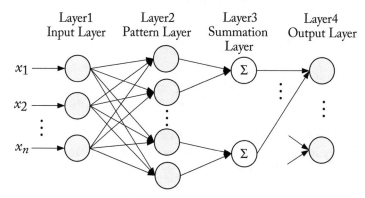

Figure 3.4: **PNN** structure.

Figure 3.4 illustrates that a PNN has four layers. In the pattern layer, the number of neurons is equal to the total patterns. The output of this layer is calculated using a nonlinear activation function, which can be described by

$$\Phi\left(I_i\right) = e^{\frac{I_j - 1}{\sigma^2}}, \tag{3.8}$$

where I_i is the weight of this layer, and σ is a parameter corresponding to the smoothness of the function. In the summation layer, all inputs are summed and the results are regarded as the final probability. Note that the number of outputs generated by the PNN is equal to the number of patterns. Additionally, all output values are binary where the single pattern with the highest probability is equal to 1.

3.4   SIMULATION SETTINGS AND RESULTS

This section introduces simulation settings, block diagrams, and simulation results for the proposed algorithm.

3.4.1   SIMULATION SETTINGS

In the simulation section, the most representative semi-active suspension control strategy, namely skyhook control, is applied to illustrate the road classifier performance for advanced suspension systems. Further, the high-fidelity Carsim model is used for evaluation. The overall strategy is described in Figure 3.5.

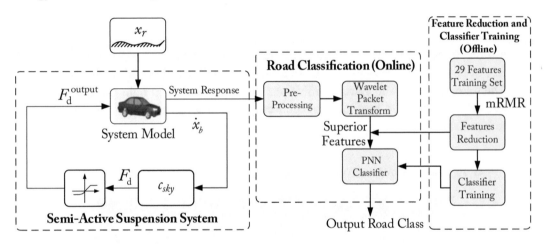

Figure 3.5: Road classification with PNN.

The strategy in Figure 3.5 contains three parts, namely, the semi-active suspension system, road classification, and classifier generation. The classifier is trained offline, and the developed classifier along with the superior features is used online.

In the classifier generation part, all 29 candidate features are first treated as inputs of the mRMR, and the generated superior features are then used to train the PNN classifier. All candidate features and their index numbers are tabulated in Table 3.3. Note that the number of the superior features is arbitrary, and more superior features means that more information is available from the system response for classifier training. However, a large number of superior features will dramatically increase classifier training time. Therefore, the operator should find a balance between computational burden and available information. In this book, the number of the superior features is set to four after some trial and error. The selected superior features and the trained PNN classifier are then pre-stored in the ECU for online application.

In the online phase, a skyhook controller is applied and generates a controllable damping force to mitigate system vibration. The measured system responses are then used for the road

Table 3.3: Index number of candidate features

Feature	No.	Description	No.	Description
Sprung mass acceleration	2	SRA of time domain	3	RMS of time domain
	12	SRA of $aD2$	23	SRA of $dA3$
Unsprung mass acceleration	2	SRA of time domain	4	Max of time domain
	20	RMS of $aD4$	29	Max of $dA3$
Rattle space	2	SRA of time domain	18	RMS of $aD2$
	20	RMS of $aD4$	23	RMS of $dA3$

classification. In this chapter, the system velocity is assumed to be a constant and known in advance. An observer or GPS can be used to estimate vehicle speed in real world applications.

The following road and system definitions are implemented to train the PNN classifier. For training purposes, road profiles corresponding to ISO levels from A–F are generated in advance (see Table 3.4).

Table 3.4: System definitions and training signal

Definition	Values
c_{sky}	1,500 Ns/m
Classification interval	1 s
Overlap	50%
Vehicle speed	40 km/h
Training road setting	100 s for each level, 600 s in total
Training road sequence	ISO-A, B, C, D, E, F
Validation road setting	50 s for each level, 300 s in total
Validation road sequence	ISO-A, D, C, F, B, A
System responses	SMA, UMA, RS

For the generated roads in either the training or validation sets, the statistical characteristics are assumed to be unchanged for each level, and the translation time for two adjacent levels is negligible.

3.4.2   SIMULATION RESULTS

Here, the simulation results for different conditions are first presented. Subsequently, a discussion of the results is carried out. Note that the classifier used in this part is trained offline based on the nominal model, and different model conditions are applied in the validation process.

Case 1. Nominal model.   In this part, the validation model has no uncertainties and all parameters are the same as those in the nominal model. The classification results are described in Figure 3.6.

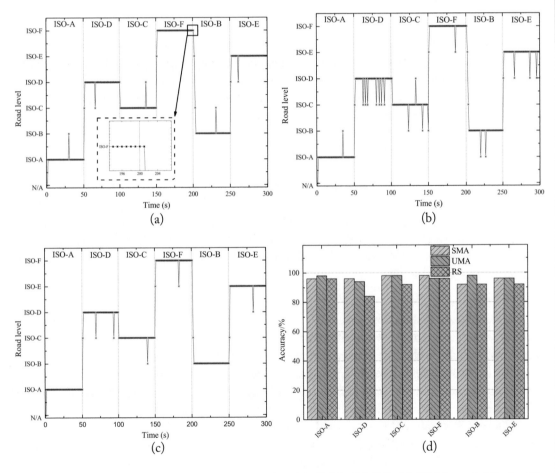

Figure 3.6: Nominal model classification results: (a) SMA, (b) RS, (c) UMA, and (d) classification accuracy.

We can see that the classification accuracy for each road level with the nominal model is more than 84%, and the minimal accuracy for SMA and UMA ARE 92% and 94%, respectively.

This means that the classification results with these two responses are better than that of RS. Apart from RS, the errors in SMA and UMA mainly appear when the road excitation level was suddenly changed. The reason for such errors can be interpreted as a result of the framing used during pre-processing. Since the overlap is set to be 50%, half of the response signal is from the previous second, this leads to aliasing and classification errors. For both SMA and UMA, the difference in classification accuracy is insignificant. The classifier based on UMA outperforms SMA for ISO-A and B, whereas the SMA classifier has higher accuracy for ISO-D. Based on this information, the conclusion can be drawn that both classifiers based on SMA and UMA perform better than that of RS for the nominal case.

Case 2. Simulation results for $c_{sky} = 4000$ Ns/m. The contradiction existing between ride comfort and handling capacity restricts suspension performance improvement. One way to remedy this is to adaptively change suspension controller gains for different road conditions. Previous research has revealed that such changes could result in up to 20% performance variation for semi-active suspension systems. In this section, the controller gain of the validation model is set to $c_{sky} = 4000$ Ns/m to test the robustness of the algorithm. This value is approximately equal to the maximum force output capacity of the adopted controllable damper and is therefore an upper limit. The results are shown in Figure 3.7.

The results in Figure 3.7 reveal that the performance of the well-trained classifiers on the new validation model is worse than that of the nominal one. However, the inaccuracies for both SMA and UMA are insignificant. For the RS-based classifier, the classification accuracies are less than 90% for road levels C, F, B, and E. Further study reveals that the controller parameter variation results in about 10% variation in the RMS of the responses. In this case, the effect can be neglected for the classifiers.

Case 3. Simulation results for $m_b = 500$ kg and $m_b = 300$ kg. Vehicle mass can vary significantly when traveling in real-world situations. Such variation can be up to 30%. Some examples of reasons for realistic changes in mass include variations in the number of passengers and consumption of gasoline. In this section, two cases where $m_b = 500$ kg and where $m_b = 300$ kg are considered to assess the robustness of the proposed algorithm. The classification accuracy of the well-trained classifiers on the two validation models are shown in Figure 3.8.

Figure 3.8 reveals that the proposed classifiers are more sensitive to sprung mass variation than variation of the controller gain. Among the three classifiers, its clear that the SMA-based classifier accuracy deteriorates the most when compared to the results of the nominal classifier. For the SMA classifier, the lowest classification accuracy for both cases reduced to 86% and 84%, while the lowest classification accuracy for the nominal model was 96%. In contrast, the lowest accuracies for the UMA-based classifier were 90% and 88%. Thus, the conclusion can be made that the UMA-based classifier is more robust to the sprung mass variation compared to the SMA-based classifier.

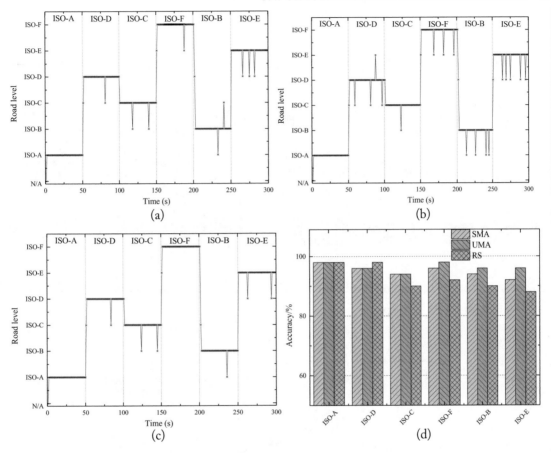

Figure 3.7: Road classification results for model with $c_{sky} = 4000$: (a) SMA, (b) RS, (c) UMA, and (d) classification accuracy.

Case 4. Further analysis in the frequency domain.   The algorithm introduced in this chapter is a time and frequency domain-based classifier. This means that information from both time and frequency domains are taken into consideration during the classifier training process. In order to show the frequency responses of the three systems, a comparison of the aforementioned conditions in the frequency domain are graphically represented in Figures 3.9–3.11. Note that the number of superior features is set to 4, and only superior features in the frequency domain are depicted in these figures. The time domain superior features, on the other hand, represent the signal energy, and the area of the shadows in the following figures can be viewed as a statistical description of these features.

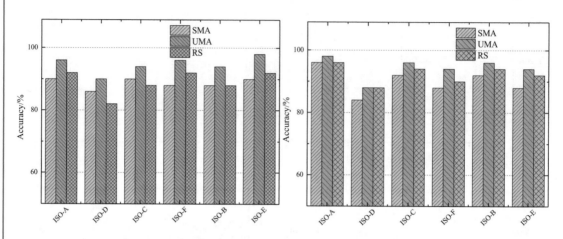

Figure 3.8: Road classification results for model with $m_b = 500$ kg and $m_b = 300$ kg: (a) $m_b = 300$ kg and (b) $m_b = 500$ kg.

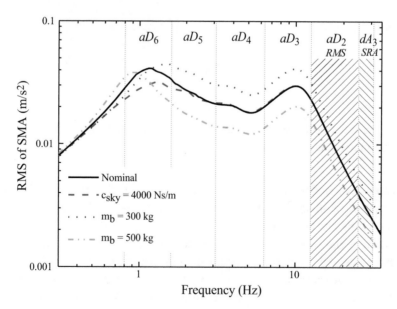

Figure 3.9: SMA frequency response.

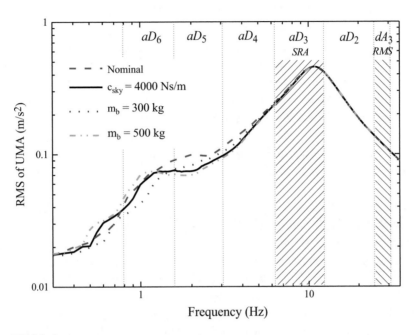

Figure 3.10: UMA frequency response.

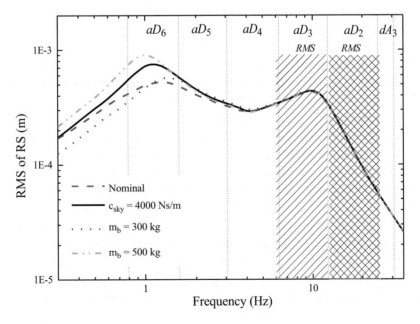

Figure 3.11: RS frequency response.

Figure 3.9 reveals that the RMS and SRA of the signals in the $aD2$ and $dA3$ regions are regarded as the superior features for the SMA-based classifier. From the results, the variation of controller gain mainly influences the SMA response in the frequency range of 0.8–3 Hz and has little effect at high frequencies. This can be interpreted as showing that the superior features in the frequency domain are relatively robust to such variations in controller gain. For the cases of $m_b = 500$ kg and $m_b = 300$ kg, we can see the sprung mass resonant frequencies shift to 0.9 Hz and 1.6 Hz, respectively. The different sprung mass will also influence the frequency response amplitude, especially in the high-frequency ranges. This can be interpreted as the reason for the deterioration of accuracy shown in Figure 3.8. This phenomenon can explain why the sprung mass-based classifier is sensitive to changes in mass resulting in errors when the sprung mass is varied.

Figure 3.10 shows that the superior features in the frequency domain for the UMA-based classifier are the SRA of $aD3$ and the RMS of $dA3$. Unlike the SMA-based classifier, the UMA classifier is insensitive to the variation of sprung mass in both frequency ranges. This indicates that this classifier is more robust to system uncertainties and controller gain variation. This can also be observed from Figure 3.8, where the performance deterioration is insignificant when compared to the nominal validation results shown in Figure 3.6.

For the RS-based classifier, the RMS of both $aD3$ and $aD2$ are taken as the superior features in the frequency domain. It can be seen from Figure 3.11 that these ranges are relatively insensitive to changes in both controller gains or sprung mass. Here, the performance of the RS-based classifier is determined by the signal alone, and the influence of the system uncertainties is relatively small.

According to the results shown in this section, we can clearly see that the UMA-based classifier outperforms the other classifiers in terms of overall accuracy and robustness. Its nearly unchanged accuracy for both the nominal and varied sprung mass models is the biggest advantage. As well, with consistent accuracy with up to ±30% variation in sprung mass is more than sufficient for mass production vehicles. For the other models, the accuracy of the SMA-based classifier on the nominal model is better than that of the RS classifier. However, the SMA-based classifier is more sensitive to variations of sprung mass, which means it is less robust.

Another phenomenon worth further attention is shown in Figures 3.9–3.11. All of the superior features are located at frequencies greater than 6 Hz. This indicates that high-frequency features are prone to be selected by mRMR. This is caused by the selection of a 1 s classification interval. This is a relatively short period even when framing is performed—especially for the low-frequency components. The fluctuations of the statistical features in the low-frequency range causes mRMR to remove these candidate features. One advantage of such a time frequency domain-based algorithm is that it can be used for different control strategies, and better classification performance can be expected if the most sensitive frequency ranges can be found in advance of the feature reduction process.

3.4.3   COMPARISON WITH OTHER METHODS

Here, we compare the proposed algorithm with two existing methods. These methods are the BPNN-based algorithm proposed by Ngwangwa et al. [47] and the ANFIS-based method proposed by Qin et al. [102]. The reason for this selection is twofold: first, the BPNN method represents a category of road classification algorithms based on information from the time domain, and the ANIFS method is based on time-frequency domain analysis with features selected by relative distance evaluation algorithm (no mutual redundancy among features is considered).

According to the results shown in Section 3.4.2, UMA is used for classification for each of the three algorithm. As well, the three well-trained classifiers are validated for four cases similar to Section 3.4.2. The definitions of the four cases are as follows:

Case 1: nominal model;

Case 2: model with $c_{sky} = 4000$ N/m;

Case 3: model with $m_b = 500$ kg; and

Case 4: model with $m_b = 300$ kg.

The simulation results for each of the three classifiers are given and discussed below. Figure 3.12 shows that the proposed algorithm outperforms the other two algorithms with an accuracy of more than 95%. The BPNN-based method achieved 80% accuracy for the nominal model, but errors are obvious when the system uncertainties are applied to the validation model.

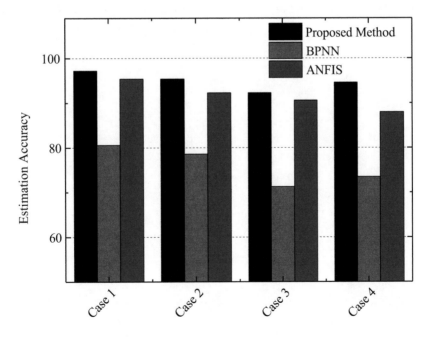

Figure 3.12: Comparison with BPNN- and ANFIS-based algorithms.

The classification accuracy decreases to 70% for the case with variable sprung mass. As for the ANFIS-based algorithm, the performance is comparable to the proposed algorithm. However, it is unable to solve the problem of repeated representation of features (i.e., data redundancy exists). In this case, the mRMR is more suitable for the issues of road classification discussed in this chapter.

3.5   SUMMARY

This chapter introduced a data-driven road classification algorithm based on measurable suspension responses. The chapter began with a comparison of system responses for various road inputs. The results showed that the differences in both time and frequency domains can be used for road classification. Next, 29 statistical features were introduced that could be used for road classification, and a feature reduction algorithm named mRMR was applied to select 4 superior features from the candidates. A PNN classifier was then used to classify road inputs into one level with the highest probability. Finally, a simulation study was carried out to evaluate the proposed algorithm, and three measurable responses, i.e., SMA, UMA, and RS, were used for classifier training. Different conditions including varied controller gains and sprung masses were applied to test the robustness of all three classifiers. From the simulation results, the UMA-based classifier performed the best for all conditions and was robust to system uncertainties. Further simulation results showed the advantage of the proposed algorithm compared to the classification algorithms that relied on time domain features or distance-based feature reduction alone.

CHAPTER 4

Model-Based Road Estimation Algorithms

This chapter introduces model-based road estimation algorithms. A transfer function-based road classification algorithm is first presented, and then an integrated observer is designed to estimate the time domain road profile.

4.1 TRANSFER FUNCTION-BASED ROAD CLASSIFICATION ALGORITHMS

This section presents a road classification method based on the system transfer function from UMA to road excitation, which is robust to speed variation. The following sections will first introduce the transfer function model and the classifier structure, and subsequently present simulation results and experimental validations.

4.1.1 STRUCTURE OF SPEED INDEPENDENT ROAD CLASSIFICATION ALGORITHM

For the model introduced in Eq. (2.39), the transfer function from UMA to road excitation is given by Eq. (4.1):

$$H(s)_{\ddot{x}_w \sim x_r} = \frac{m_b m_w s^4 + (c_p m_b + c_p m_w) s^3 + A + B + k_s k_t}{m_b k_t s^4 + c_p k_t s^3 + k_s k_t s^2}$$

$$A = (m_w k_s + m_b k_s + m_b k_t + c_p^2 - c_p^2 m_w / m_b) s^2 \tag{4.1}$$

$$B = (c_p k_t + c_p k_s - c_p k_s m_w / m_b) s.$$

As discussed in Section 2.1, the structure of real roads may vary from 1.5–3.5. Without a loss of generality, this book uses six representative road levels as shown in Table 2.2. The overall structure of the speed independent road classification algorithm is shown in Figure 4.1 [45].

The present algorithm aims to allocate the underlying road input into one of the six levels based on measurements of UMA and vehicle velocity, only. Generally, the structure of the classification algorithm contains two phases, namely, online and offline. In the offline phase, the transfer function $H(s)$ and a classifier that can classify road class are generated. In the online phase, the equivalent road profile $x_{er}(t)$ is calculated according to the unsprung mass acceleration and $H(s)$. Note that $x_{er}(t)$ is not equal to the real road profile, but it has the same statistical

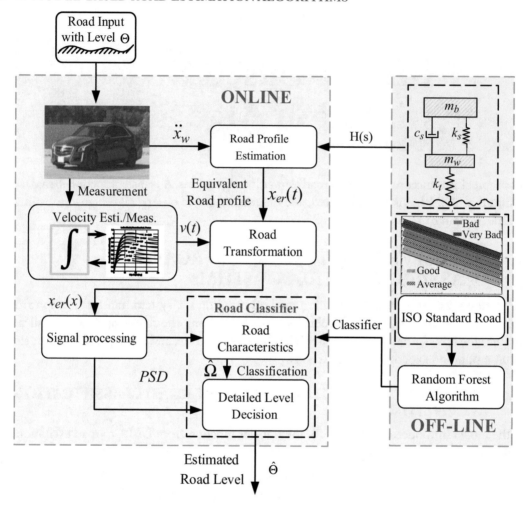

Figure 4.1: Structure of the proposed classification algorithm.

characteristics to $x_r(t)$. With the measured vehicle velocity and $x_{er}(t)$, the equivalent road profile $x_{er}(x)$, which is a function of road length x, can then be calculated. $x_{er}(x)$ is then processed using a low-pass filter and framing, and the corresponding PSD is then calculated. A two-step classifier is finally used to determine the detailed road level.

The two-step classifier first provides a rough road estimation, and then attributes the smoothed PSD into one level depending on the waviness. To provide a rough estimation at the first step, we use the ISO road definition, and merge two adjoining road levels into a single new road class: ISO-A and ISO-B become the "Good" class, ISO-C and ISO-D become the "Average" class, ISO-E and ISO-F become the "Poor" class, and the remaining two levels become the "Very Poor" class. The class is determined by six center frequencies ranging from 0.0625–2 (cycles/m), and the detailed definition is shown in Figure 4.2 [45].

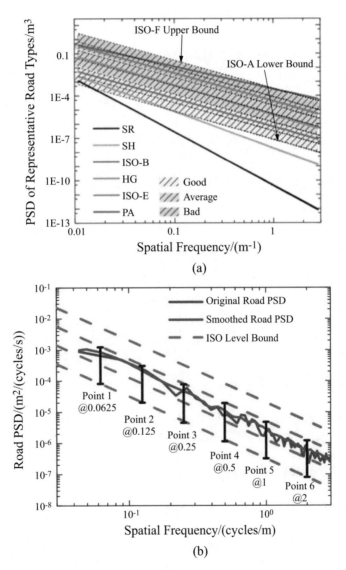

Figure 4.2: Road class: (a) road classes and six levels and (b) determination of road class.

In Figure 4.2a, we can see that SR, SH, and ISO-B are located inside the "Good" class, HG belongs to the "Average" class, and both ISO-E and PA are in the "Bad" class. Further, Figure 4.2b shows the original estimated PSD and illustrates how to determine road class based on the smoothed PSD. The smoothed PSD can be obtained by following the method given in

ISO 8608 [106]. The method in ISO 8608 is a threshold-based algorithm, and the bounds of individual ISO level at the six central frequencies are tabulated in Table 4.1.

Table 4.1: Central frequencies of different road levels [106]

Road Level	$G_q(n_c)$ (10⁻⁶ m³)	Point (Central frequency) (cycles/m)					
		1 (0.0625)	2 (0.125)	3 (0.25)	4 (0.5)	5 (1)	6 (2)
A	Mean Value	41	10.2	2.56	0.64	0.16	0.04
	Upper Bound	81.9	20.5	5.12	1.28	0.32	0.08
B	Lower Bound	81.9	20.5	5.12	1.28	0.32	0.08
	Mean	163.8	41	10.24	2.56	0.64	0.16
	Upper Bound	327.7	81.9	20.48	5.12	1.28	0.32
C	Lower Bound	327.7	81.9	20.48	5.12	1.28	0.32
	Mean	655.4	163.8	40.96	10.24	2.56	0.64
	Upper Bound	1310.7	327.7	81.92	20.48	5.12	1.28
D	Lower Bound	1310.7	327.7	81.92	20.48	5.12	1.28
	Mean	2621.4	655.4	163.84	40.96	10.24	2.56
	Upper Bound	5242.9	1310.7	327.68	81.92	20.48	5.12
E	Lower Bound	5242.9	1310.7	327.68	81.92	20.48	5.12
	Mean	10485.8	2621.4	655.36	163.84	40.96	10.24
	Upper bound	20971.5	5242.9	1310.72	327.68	81.92	20.48

For a road to be classified, all of the results determined by Table 4.1 are then taken as the input of a random forest classifier [107] and the road is then attributed to one of the four classes.

Once the road class is determined, the second step is to decide the road level. A procedure shown in Table 4.2 is used for this purpose. The main idea of the procedure is to use the linear fit of waviness resulting from the smoothed PSD to determine the road level. Note that no prior training process is required in the proposed algorithm, which is a great competitive advantage to traditional data-driven algorithms.

An example for the proposed algorithm is discussed for road level ISO-B, with a road length of 400 m at speed of 36 km/h. The classification results for an interval of 20 m are given below. We can see from Figure 4.3a that the classification accuracy is 90%, with errors appearing at 200 m and 340 m. With the exception of these two points, the result in Figure 4.3b shows that although the random forest classifier attributes some of the frequency points to wrong class, (e.g., 40 m, 100 m, and 240 m), more points are correctly classified. Figure 4.3c shows the final road level estimation results based on the fitted waviness. Since all the results are bigger than −2.3 with a mean value of −2.06, the proposed algorithm can provide road level classification

Table 4.2: Road level determination procedure

If $\hat{\Omega}$ is ***Class Good***, then
   If $|\hat{W}| \geq 2.8$, then
      Road level ***SR*** detected
   Else if $2.3 \leq |\hat{W}| \leq 2.8$, then
      Road level ***SH*** detected
   Else if $|\hat{W}| \leq 2.3$, then
      Road level ***ISO-B*** detected
Else if $\hat{\Omega}$ is ***Class Average***, then
   Road level ***HG*** detected
Else if $\hat{\Omega}$ is ***Class Bad***, then
   If $|\hat{W}| \geq 1.8$, then
      Road level ***ISO-E*** detected
   Else if $|\hat{W}| \leq 1.8$, then
      Road level ***PA*** detected
Else
   Road level ***Unknown*** detected
End

results for a constant vehicle speed. More simulations are presented next to show the robustness of the present algorithm.

4.1.2 DATA PROCESSING PROCEDURE

Since the road profiles in this chapter are randomly generated, the following steps are performed in the following sections to avoid possible statistical errors.

Step 1. Make a 6,000 m length road containing all road levels defined in Table 2.2, each with a length of 1,000 m.

Step 2. Calculate $x_{er}(x)$ based on the vehicle velocity and the sampled unsprung mass acceleration signal.

Step 3. Apply a bootstrapping method to pick random segments for classification. Each segment length is 50 and is resampled 5 times.

Step 4. Use the proposed algorithm, and use the mean value of all the re-sampling results to assess the overall accuracy.

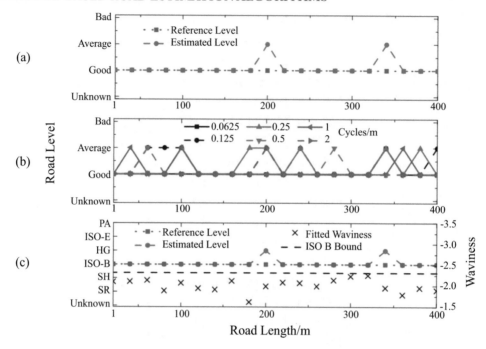

Figure 4.3: Classification result for the sample road: (a) road classification results, (b) estimated road class for each center frequency, and (c) road level and waviness.

4.1.3   DISCUSSION ON THE INPUT SELECTION OF RF CLASSIFIER

Figure 4.3a presents the road classification results for an example period of an ISO-B road where the road class was determined using a RF classifier. Based on the road class calculated using individual center spatial frequencies, the RF classifier outputs the estimated road class. RF is composed of many decision trees, with individual classifiers generated by random vectors that are separately sampled from the system input. Each decision tree of the well-trained classifier contributes a single vote to the allocation of the input into a level with the highest probability.

For the results shown in Figure 4.3a, we can notice that, the results at 0.125, 0.25, and 2 cycles/m are classified as part of the "Average" class at 200 m, whereas the results of other center spatial frequencies are classified as "Good." Since both classes contain the same quantity of points, the well-trained RF classifier attributes this road period to the "Average" class, and the result at 340 m is the same. It is of note that for the result at 100 m, although both classes contain the same center frequency points, the classifier attributes this single part to "Good" class. Since all other segments are classified to the class that contains the highest number of center spatial frequencies points, we hypothesize that the well-trained classifier will provide random estimation to either road class when these two classes have the same quantity of points. To prove this hypothesis, we generate a matrix with dimensions of 100 × 6. For each row, we then randomly select three elements and set their values to "Good," with the other three elements set as "Average." The generated matrix is used as the input to the generated RF classifier. After 50 runs, 50.5% of the outputs are classified as "Average." The result of one example run is shown in Figure 4.4.

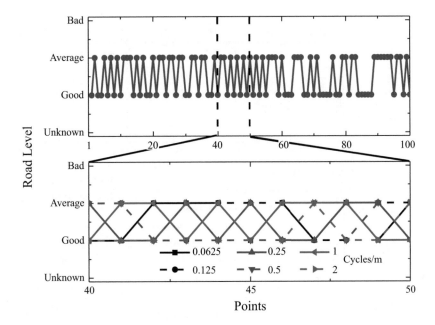

Figure 4.4: 100 points classification result.

We can see from Figure 4.4 that for the two classes that contain the same quantity of points the trained RF classifier will output either of the two classes. One resolution to this issue is to change the number of inputs to the classifier to an odd integer by either adding or removing one or more points from the input set.

4.1.4   SIMULATION RESULTS FOR VARYING VELOCITY SCENARIO

Since vehicle speed rapidly changes due to traffic conditions and speed limits, it's important to test the robustness of a road classification algorithm with respect to varying velocity. This section presents simulation results of the proposed algorithm in a varying velocity scenario. The velocity map is shown in Figure 4.5a, and contains two well-known driving cycles from the EPA inspection and maintenance standard (IM240) and the highway fuel economy driving schedule (HWFET). The classification results are shown in Figure 4.5 and Table 4.3. Note that the classification interval for this scenario is 20 m with 10 m of information from the previous measurement. This results in a void input at 10 m and 99 output points for PA. We can see from Table 4.3 that the classification accuracy is more than 75% except for the SH. The reason for this is that in Table 4.2, the SH road level is determined by both the upper and lower bounds, and is therefore stricter than the other road levels. According to the results given in Figure 4.5 and Table 4.3, we can come to the conclusion that the proposed algorithm is robust to vehicle speed variation.

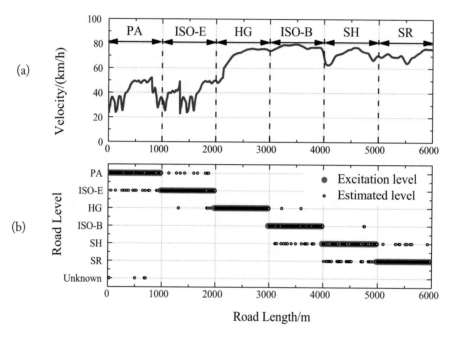

Figure 4.5: Classification result for varying velocity scenario: (a) velocity map and (b) classification result.

Table 4.3: Classification accuracy for varying velocity

Road Level		Classified Level							Accuracy
		PA	ISO-E	HG	ISO-B	SH	SR	Unknown	
True Level	PA	76	18	–	–	–	–	5	76.7%
	ISO-E	8	88	4	–	–	–	–	88%
	HG	–	–	100	–	–	–	–	100%
	ISO-B	–	–	2	81	17	–	–	81%
	SH	–	–	–	6	64	30	–	64%
	SR	–	–	–	–	9	91	–	91%

4.1.5   SIMULATION RESULTS FOR NOISY MEASUREMENT

All simulation results discussed above have used responses without measurement noise. However, in reality, measurement noise is unavoidable. This section investigates the influence of noise on the estimated PSD and classification accuracy. First, we define the signal-to-noise ratio (SNR) as follows:

$$SNR = \frac{P_{signal}}{P_{noise}} = \frac{\sigma^2_{signal}}{\sigma^2_{noise}}, \tag{4.2}$$

where σ^2 stands for signal variance. For the purposes of this section, we set the SNR to be 10, 20, and 40. The influence of these noise levels is shown in Figure 4.6. The red area of each subplot is the result of 20 runs with different numbers of seeds. Figure 4.6 shows that noise level variation mainly influences the low-frequency range of the estimated PSD. In this low-frequency range, larger noise results in more significant overestimation. In contrast, the high-frequency range of the estimated PSD is insensitive to measurement noise. Note that for $SNR = 10$, the estimated PSD values at a center frequency of 0.0625 cycles/m is increased "Average" class. This indicates that it will be helpful to improve classification accuracy by removing this point from the classifier input.

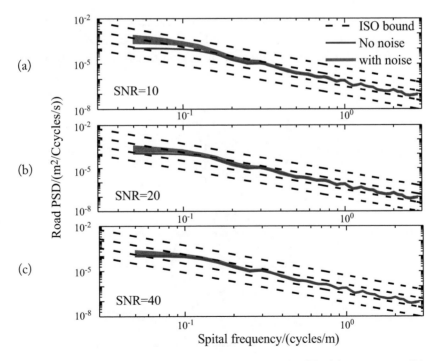

Figure 4.6: Influence of measurement noise on the estimated PSD: (a) $SNR = 10$, (b) $SNR = 20$, and (c) $SNR = 40$.

Considering the discussion in Section 4.1.3, we suggest using five center spatial points to form the input by abandoning the center spatial point at 0.0625 cycles/m.

4.1.6   ROAD CLASSIFICATION WHEN CONSIDERING TIRE DYNAMICS

Another important factor that may affect vehicle vertical response is tire enveloping. In the following section, we will develop a tire enveloping model, and show the classification results using this model along with the proposed algorithm.

Tire enveloping model.   The single-point contact assumption used in Eq. (2.39) neglects tire enveloping effects and may overestimate road input. To acquire a more realistic system response, the following section introduces the generation of the effective road profile using the flexible roller contact (FRC) theory [108].

For a tire rolling over road profile shown in Figure 4.7, the compressing deformation between the road profile and tire can be defined as

$$
\begin{aligned}
x_{td}(y, \Delta) &= [h(y, \Delta) - h_O(0)] - x_w(y) \\
&= x_r(y + \Delta) + \sqrt{R^2 - \Delta^2} - R - x_w(y),
\end{aligned}
\tag{4.3}
$$

where y is the road longitudinal coordinate and Δ is local coordinate associated with the tire center O. $h(y, \Delta)$ is the tire center height at (y, Δ). The tire center height at the origin point is denoted by $h_O(0)$, and R is the tire radius. Considering the one-sided constraint property of

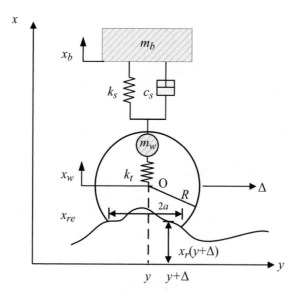

Figure 4.7: Geometry of tire rolling over road profile.

road profiles, the actual tire compression deformation x_{td}^* is defined as

$$x_{td}^* (y, \Delta) = \begin{cases} x_{td} (y, \Delta), & x_{td} \geq 0 \\ 0, & x_{td} < 0. \end{cases} \tag{4.4}$$

With Eq. (4.3), the vertical tire force can be calculated by

$$F_{zd} (y) = \int_{-a}^{a} k_{te} (\Delta) \cdot x_{td}^* (y, \Delta) \, d\Delta, \tag{4.5}$$

where k_{te} is the tire vertical stiffness coefficient defined as follows:

$$k_{te} (\Delta) = \frac{k_t}{2a} \tag{4.6}$$

and a is the half-contact length. When a vehicle travels with a small enough velocity, the equivalent road profile $x_{re} (y)$ is equal to the unsprung mass displacement $x_w (y)$. $x_{re} (y)$ can be calculated using

$$\begin{cases} \int_{-a}^{a} c_t (\Delta) \left[x_r (y + \Delta) - \sqrt{R^2 - \Delta^2} - R - x_{re} (y) \right] d\Delta = F_{ts} \\ x_r (y + \Delta) - \sqrt{R^2 - \Delta^2} - R - x_{re} (y) \geq 0 \\ -a \leq \Delta \leq a. \end{cases} \tag{4.7}$$

For an ISO-B road, the FRC model is then applied to generate the effective road profile, and a comparison between the effective and original road profiles is shown in Figure 4.8. It can be seen in Figure 4.8a that the detailed component of the effective road is filtered out and the low-frequency trend remains unchanged. Comparing the results show that the effective road profile performs the same as the original road in low-frequency range. When the spatial frequency is larger than 0.7 cycles/m, the effective road profile PSD continues decreasing with increasing spatial frequency.

Road classification. With the sampled unsprung mass acceleration in the time domain, frequency characteristics of input road can be calculated via the frequency response function defined by Eq. (4.8):

$$|H (\Omega)|^2_{x_r \sim \ddot{x}_w} = \frac{G_{\ddot{x}_w} (\Omega)}{G_{x_r} (\Omega)}. \tag{4.8}$$

The transfer function shown in Eq. (4.8) can be obtained for HG, which is shown in Figure 4.9 including tire enveloping effects.

The proposed algorithm is then validated based on the obtained transfer function. For the 400 m sample road used in Section 4.1.1, the classification result is shown in Figure 4.10. We can see from Figure 4.10 that the classification accuracy is 95%. This means that 19 road segments were accurately estimated. In Figure 4.10a, the road section at 60 m is misclassified

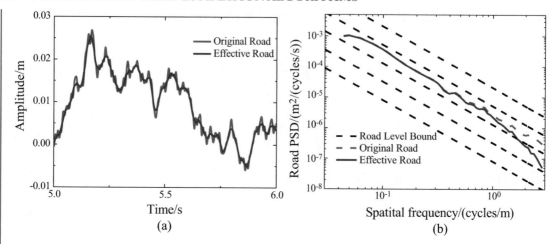

Figure 4.8: ISO-B, 40 km/h road profile with tire enveloping effect: (a) time-domain comparison and (b) PSD comparison.

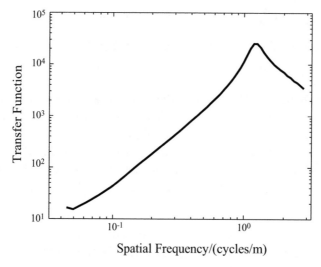

Figure 4.9: Frequency response function for unsprung mass acceleration.

to "Average." This results in one classification error point in Figure 4.10c. Using the simulation results, we can conclude that the proposed algorithm is capable of providing satisfactory road classification when the tire dynamics are considered. Additionally, the transfer function from the unsprung mass acceleration to the road excitation can be used to provide satisfactory input before the classification procedure.

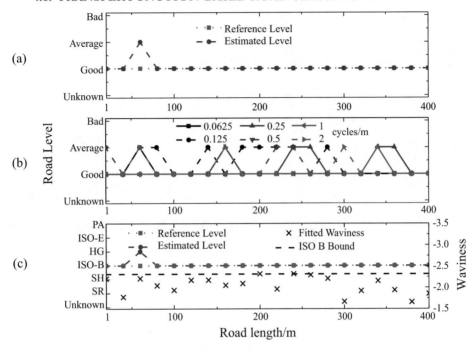

Figure 4.10: Classification results considering tire enveloping: (a) estimated class, (b) estimated class at center frequencies, and (c) estimated road level.

4.1.7   EXPERIMENTAL VALIDATION

This section presents the experimental validation for the proposed algorithm. The test field is composed of pavement and gravel type roads, located in the Waterloo region of Ontario, Canada. The test field is shown in Figure 4.11.

A Cadillac CTS was taken as the test vehicle, and the transfer function model was calculated according to the nominal parameters. This means that no training process was performed prior to the following tests. The unsprung mass acceleration was measured by an accelerometer (Kistler 8305), and vehicle speed was calculated according to a tachometer on the wheel hub. The detailed data processing procedure is described in Figure 4.12. The measurements were measured and processed using a dSpace Micro Autobox II, and the algorithm used the processed data to estimate road level. A bandpass filter with cut-off frequencies of 0.3 Hz and 100 Hz was adopted to filter the measurements [45].

In order to evaluate the robustness of the proposed algorithm, three velocities, i.e., 20, 50, and 80 km/h, were tested, and the number of passengers was one or four. The test conditions are tabulated in Table 4.4.

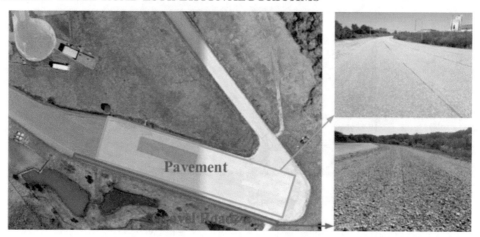

Figure 4.11: Test field with gravel and pavement road (image courtesy of Google map).

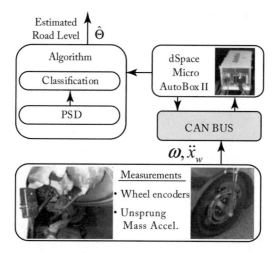

Figure 4.12: Data processing procedure.

The results of test No. 7 are shown in Figure 4.13, and the vehicle speed shown in Figure 4.13a was calculated according to the wheel speed and wheel radius. Based on the transfer function and the filtered unsprung mass acceleration in Figure 4.13b, the equivalent road profile was calculated and is displayed in Figure 4.13c. Note that the equivalent road profile did not contain any vehicle speed information and had almost the same statistical characteristics to the real road excitation.

Based on the above discussion, the final classification results are shown in Table 4.5. The test conditions in Table 4.4 were divided into three groups, namely, more passengers in good

Table 4.4: Test conditions

No.	Road Type	Speed (km/h)	Passenger Number	Duration (s)	Excitation Level
1	Pavement	20	4	20	SH
2	Pavement	50	4	6	SH
3	Pavement	80	4	21	SH
4	Pavement	20	1	17	SH
5	Pavement	50	1	8	SH
6	Pavement	80	1	5	SH
7	Gravel	20	4	21	HG
8	Gravel	50	4	12	HG
9	Gravel	80	4	4	HG

Table 4.5: Experimental validation results

Combination	SH	ISO-B	HG	ISO-E	Actual Excitation Level	Accuracy
1 (Test 1-3)	53	11	1	–	SH	81%
2 (Test 4-6)	19	5	1	–	SH	76%
3 (Test 7-9)	–	3	30	–	HG	91%

road conditions, nominal passenger number in good road conditions, and more passengers in bad road conditions. It can be seen from Table 4.4 that the accuracy of combination 3 is more than 90%, and only 3 samples are mislabeled into ISO-B. As for combinations 1 and 2, the overall accuracy exceeds 75% and most of the incorrect results were classified to ISO-B, with only one segment labeled as HG. This indicates that the first RF classifier can accurately attribute road class, but the waviness-based road level determination procedure is less accurate. Generally, the classification accuracy for combinations 1 and 2 is good. Therefore, the conclusion can be drawn that the proposed algorithm is robust to parameter variations such as vehicle velocity and passenger number.

4.2   OBSERVER-BASED ROAD PROFILE ESTIMATION

This section introduces a novel road estimation algorithm based on an observer structure. As introduced in Chapter 1, the traditional unknown road observers were developed based on the assumption that all variables were measurable, which is difficult to achieve in practice. The novel observer structure in this section uses both sprung and unsprung mass accelerations which can

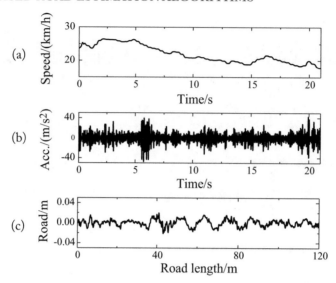

Figure 4.13: Experiment result for test No. 7: (a) vehicle speed, (b) unsprung mass acceleration, and (c) calculated equivalent road profile.

be easily measured with commercially available sensors. First, the observer structure is presented, and then simulation results are provided.

4.2.1   OBSERVER STRUCTURE

The proposed observer is composed of two observers in series, namely, an adaptive Kalman filter (AKF) and an adaptive super twisting observer (ASTO). The overall structure of the proposed AKF-ASTO observer is shown in Figure 4.14. In order to demonstrate the effectiveness of the proposed algorithm on controllable suspension systems, a simple yet classical semi-active suspension controller, a skyhook control, is applied.

As shown in Figure 4.14, the proposed AKF-ASTO is composed of three main components. In the AKF part, an AKF is adopted and its noise covariance matrices were tuned using the estimated road level. Measurements are taken of sprung mass and unsprung mass accelerations, and the method mentioned in Chapter 3 can be used for road level estimation. The estimated sprung mass and unsprung mass velocities are then used by the other components. In the control strategy part of the model, the control law calculates the ideal controllable damper force. The real applied force is the saturated force generated by the boundary damper model discussed in Section 2.2.3. The estimated unsprung mass velocity and controllable damper force are then used in the ASTO part of model. Finally, the ASTO outputs the estimated road profile in the time domain.

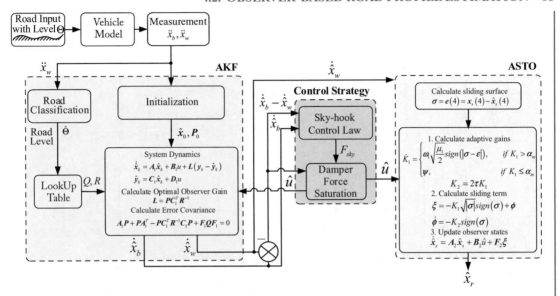

Figure 4.14: Overall structure of AKF-ASTO.

4.2.2 AKF-ASTO DESIGN

This section introduces the design of the AKF and ASTO observers.

AKF The purpose of AKF is to generate the optimal solution of an optimization problem based on a system model with varying processes and the measurement noise covariance matrices. The following three requirements must be satisfied when designing the observer:

- minimize the influence of the measurement and process noise;

- provide the estimated states for the skyhook control; and

- avoid drifting in the estimation of $\dot{x}_w$ due to the DC-offsets.

For the AKF, the system state and state are designed as $\mathbf{x}_k = [x_b - x_w, x_w - x_r, \dot{x}_b, \dot{x}_w]^T$ and $\mathbf{y}_k = [\ddot{x}_b, \ddot{x}_w]^T$. The state space equations for the system shown in Eq. (5.24) can be described by:

$$\dot{\mathbf{x}}_k = \mathbf{A}_1\mathbf{x}_k + \mathbf{B}_1\hat{u} + \mathbf{F}_1 w$$
$$\mathbf{y}_k = \mathbf{C}_1\mathbf{x}_k + \mathbf{D}_1\hat{u} + v, \qquad (4.9)$$

where both the process noise w and measurement noise v satisfy the following conditions:

$$E(w) = E(v) = 0, \quad E(ww^T) = Q, \quad E(vv^T) = R, \quad E(wv^T) = 0.$$

The system matrices can be described by:

$$
A_1 = \begin{bmatrix} 0 & 0 & 1 & -1 \\ 0 & 0 & 0 & 1 \\ -\dfrac{k_s}{m_b} & 0 & 0 & 0 \\ \dfrac{k_s}{m_w} & -\dfrac{k_t}{m_w} & 0 & 0 \end{bmatrix}, \qquad B_1 = \begin{bmatrix} 0 \\ 0 \\ -\dfrac{1}{m_b} \\ \dfrac{1}{m_w} \end{bmatrix},
$$

$$
F_1 = \begin{bmatrix} 0 \\ -1 \\ 0 \\ 0 \end{bmatrix}, \quad C_1 = \begin{bmatrix} -\dfrac{k_s}{m_b} & 0 & 0 & 0 \\ \dfrac{k_s}{m_w} & -\dfrac{k_t}{m_w} & 0 & 0 \end{bmatrix}, \quad \text{and} \quad D_1 = \begin{bmatrix} -\dfrac{1}{m_b} \\ \dfrac{1}{m_w} \end{bmatrix}.
$$

It is seen that the observability matrix is of full rank. We can then design a KF as follows:

$$
\begin{aligned}
\dot{\hat{x}}_k &= A_1\hat{x}_k + B_1\hat{u} + L\left(y_k - \hat{y}_k\right) \\
\hat{y}_k &= C_1\hat{x}_k + D_1\hat{u},
\end{aligned}
\tag{4.10}
$$

where L is the optimal feedback gain, which can be described by

$$
L = PC_1^T R^{-1},
\tag{4.11}
$$

and P is the solution to the following Riccati equation

$$
A_1 P + PA_1^T - PC_1^T R^{-1} C_1 P + F_1 QF_1 = 0.
\tag{4.12}
$$

Note that a previous study has shown that improper Q and R matrices can dramatically degrade the observer performance [54]. Fortunately, these values are proportional to the road excitation level. In this case, we can use the road classification method discussed in Chapter 3 to first determine road excitation level, and then find Q and R using a look-up table. The values of Q and R for different road levels can also be pre-calculated.

Using the road definition given in Eq. (2.1), we can write the road velocity PSD in the spatial frequency domain as follows:

$$
G_{\dot{q}}(n_s) = (2\pi n_0)^2 G_q(n_0).
\tag{4.13}
$$

Here, we can see that all parameters in the above equation are constants. This means that the road velocity is a white noise. By setting the spatial frequency bounds to be 0.011 m^{-1} and 2.83 m^{-1}, the process noise Q for different ISO levels can be calculated.

As for the measurement noise covariance R, we can use the assumption that it is proportional to the variance of the acceleration signal, and the SNR ratio is set to be 0.05%. Both Q and R values are tabulated in Table 4.6.

Table 4.6: $\mathbf{Q}$ and $\mathbf{R}$ values for different road levels

Road Class	Q (m²/s²)	R (m²/s⁴)	
		Sprung Mass	Unsprung Mass
A	0.0022	2.5 × 10⁻⁵	4 × 10⁻⁴
B	0.0088	1 × 10⁻⁴	1.6 × 10⁻³
C	0.0351	4 × 10⁻⁴	6.4 × 10⁻³
D	0.1406	1.6 × 10⁻³	2.56 × 10⁻²
E	0.5624	6.4 × 10⁻³	1.024 × 10⁻¹
F	2.2495	2.56 × 10⁻²	4.096 × 10⁻¹

ASTO This section uses the ASTO proposed by Shtessel et al. to estimate the unknown road excitation in the time domain [109]. The biggest advantage of ASTO is that it can adaptively tune the observer gains and requires no prior information of the unknown input. Selecting the new system states as $\mathbf{x}_s = [x_b, \dot{x}_b, x_w, \dot{x}_w]^T$, the state space equation of the designed ASTO is described by

$$\dot{\mathbf{x}}_s = \mathbf{A}_2\mathbf{x}_s + \mathbf{B}_2\hat{u} + \mathbf{F}_2 x_r, \tag{4.14}$$

where the matrices are

$$\mathbf{A}_2 = \begin{bmatrix} 0 & 1 & 0 & 0 \\ -\dfrac{k_s}{m_b} & 0 & \dfrac{k_s}{m_b} & 0 \\ 0 & 0 & 0 & 1 \\ \dfrac{k_s}{m_w} & 0 & -\dfrac{k_t + k_s}{m_w} & 0 \end{bmatrix}, \quad \mathbf{B}_2 = \begin{bmatrix} 0 \\ 1 \\ -\dfrac{1}{m_b} \\ 0 \\ \dfrac{1}{m_w} \end{bmatrix}, \quad \mathbf{F}_2 = \begin{bmatrix} 0 \\ 0 \\ 0 \\ \dfrac{k_t}{m_w} \end{bmatrix}.$$

The ASTO is then designed to observe the unknown road excitation:

$$\dot{\hat{\mathbf{x}}}_s = \mathbf{A}_2\hat{\mathbf{x}}_s + \mathbf{B}_2\hat{u} + \mathbf{F}_2\xi, \tag{4.15}$$

where ξ is a sliding term determined by the following adaptive super twisting laws:

$$\xi = -K_1\sqrt{|\sigma|}\,\text{sign}\,(\sigma) + \phi$$
$$\dot{\phi} = -K_2\text{sign}\,(\sigma), \tag{4.16}$$

where the gains K_1, K_2 are adaptive gains that can be calculated by

$$\dot{K}_1 = \begin{cases} \omega_1\sqrt{\dfrac{\mu_1}{2}}\text{sign}\,(|\sigma| - \varepsilon), & \text{if } K_1 > \alpha_m \\ \psi, & \text{if } K_1 \leq \alpha_m \end{cases} \tag{4.17}$$

$$K_2 = 2\tau K_1, \tag{4.18}$$

where ω_1, ψ, τ, μ_1, and α_m are positive gains to be designed.

We then define the estimation error as $\mathbf{e} = \mathbf{x}_s - \hat{\mathbf{x}}_s$. The observer is designed to drive the error to converge to a bounded area in a finite time. The unknown road profile is then reconstructed as per the term ξ. To this end, we define the sliding surface as

$$\sigma = \mathbf{e}(4) = \mathbf{x}_s(4) - \hat{\mathbf{x}}_s(4), \tag{4.19}$$

and its time derivative can be described by

$$\begin{aligned}
\dot{\sigma} = \dot{\mathbf{e}}(4) &= \mathbf{A}_2(4,:)\,\mathbf{x}_s + \mathbf{B}_2(4,:)\,\hat{u} + \mathbf{F}_2(4,:)\,x_r \\
&\quad - [\mathbf{A}_2(4,:)\,\hat{\mathbf{x}}_s + \mathbf{B}_2(4,:)\,\hat{u} + \mathbf{F}_2(4,:)\,\xi] \\
&= \underbrace{\mathbf{A}_2(4,:)\,\mathbf{e}}_{a_1(e,t)} + \underbrace{\mathbf{F}_2(4,:)\,x_r}_{a_2(t)} \underbrace{-\mathbf{F}_2(4,:)\,\xi}_{\omega(t)} \\
&= a_1(\mathbf{e},t) + a_2(t) + \omega(t),
\end{aligned} \tag{4.20}$$

where $a_1(\mathbf{e},t)$ represents the estimation error, $a_2(t)$ is the unknown real road excitation, and $\omega(t)$ represents the dynamics of the observer. We can then design the observer by making the following assumptions.

Assumption 4.1 Estimation error $a_1(\mathbf{e},t)$ is a function with a positive constant bound of δ_1:

$$|a_1(\mathbf{e},t)| \le \delta_1\sqrt{|\sigma|}. \tag{4.21}$$

Assumption 4.2 The road excitation function $a_2(t)$ is bounded by an unknown positive constant δ_2:

$$|a_2(x_r,t)| \le \delta_2. \tag{4.22}$$

Theorem 4.3 *With Assumptions 4.1 and 4.2, the observer proposed in Eq. (4.15) can drive $\sigma, \dot{\sigma} \to 0$ in a finite time via the adaptive laws given in Eq. (4.16) in the presence of bounded disturbances.*

Proof. Consider the following Lyapunov function [109]:

$$V(s_1, s_2, K_1, K_2) = V_0 + \frac{1}{2\mu_1}\left(K_1 - K_1^*\right)^2 + \frac{1}{2\mu_2}\left(K_2 - K_2^*\right)^2, \tag{4.23}$$

where μ_2, K_1^*, K_2^* are positive constants and $K_1 - K_1^* < 0$, $K_2 - K_2^* < 0$.

The function V_0 is chosen as [110]:

$$V_0 = \mathbf{s}^T \mathbf{P} \mathbf{s}, \tag{4.24}$$

where $\mathbf{s} = [s_1\ s_2]^T = \left[\sqrt{|\sigma|}\,\text{sign}\,(\sigma)\ a_2 + \phi\right]^T$. With the Lyapunov function and the adaptive control gains, $\dot{V}\,(s_1,\ s_2,\ K_1,\ K_2)$ can be proven to be negative definite. Therefore, the convergence of σ and $\dot{\sigma}$ can be ensured and the real second-order sliding mode may be established. This is similar to the process introduced by Shtessel et al. [109]. $\square$

4.2.3   SIMULATION RESULTS

This section introduces the numerical simulation results of the proposed AKF-ASTO approach. The controller gains used here are: $\omega_1 = 50$, $\psi = 32$, $\tau = 12$, $\varepsilon = 0.5$, and $\alpha_m = 2$. These values were chosen by trial and error, and the initial conditions are set to be equal to 0. The proposed algorithm is tested on a complex road composed of various road levels including ISO-A, B, and C. The test road profile is generated at 40 km/h and is shown in Figure 4.15. Each road level in the test road lasts for 10 s, and the translation time of any two adjacent levels is neglected.

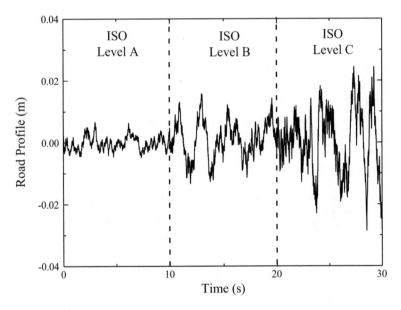

Figure 4.15: Test road profile.

The simulation results for the test road profile is shown in Figure 4.16. A traditional KF based algorithm proposed by Doumiati et al. [40] is used for comparison. This method uses KF to solve an extended-state problem, and the road profile was taken as an extra system state. Note that in this KF method, both process and measurement noise are treated as constants. Figure 4.16 shows that the present AKF-ASTO approach can accurately estimate the road inputs of the three levels, and rapidly converges to the real excitation for sudden changes in road level. In

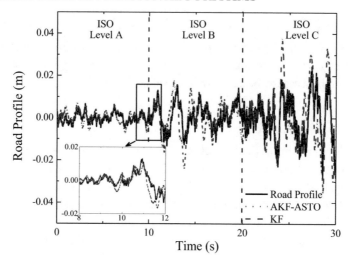

Figure 4.16: Road estimation result.

regards to the alternative KF method, although the trend is similar to that of the real excitation, the estimation errors become larger for poorer road conditions.

In order to further illustrate the influence of the noise covariance on the estimation accuracy, the STD of both the adaptive and constant covariance methods for the three road levels are shown in Table 4.7.

Table 4.7: Influence of noise covariance on estimation accuracy

Variance	STD of Estimation Error (m)		
	ISO-A	ISO-B	ISO-C
Adaptive	4.8×10^{-4}	8.8×10^{-4}	2.4×10^{-3}
ISO-A	4.8×10^{-4}	1.1×10^{-3}	6.3×10^{-3}
ISO-B	5.1×10^{-4}	8.8×10^{-4}	5.3×10^{-3}
ISO-C	5.6×10^{-4}	1.1×10^{-3}	2.4×10^{-3}

It can be seen from Table 4.7 that the proposed road classification-based adaptive algorithm has the highest accuracy. As well, the influence of improper covariance is more obvious for worse road conditions. The STD of estimation error is about 2.5 times higher than the adaptive case for road level ISO-C. Conversely, this difference for a road level of ISO-A is insignificant.

Finally, the estimated states of ASTO are shown in Figure 4.17. It can be seen that ASTO can accurately estimate all states for the three road levels. Therefore, we can conclude that the

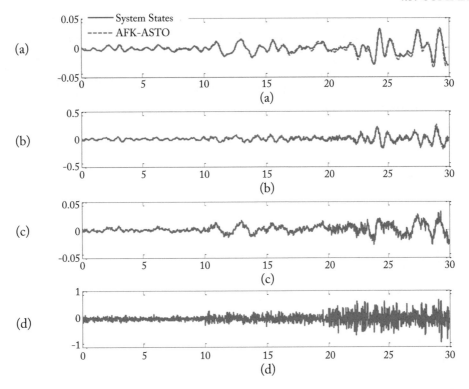

Figure 4.17: Estimated system states of ASTO: (a) displacement of sprung mass, (b) velocity of sprung mass, (c) displacement of unsprung mass, and (d) velocity of unsprung mass.

proposed AKF-ASTO can effectively estimate road profiles and system states simultaneously solely with measurements of sprung and unsprung mass accelerations.

4.3   SUMMARY

This chapter introduced model-based road estimation algorithms. Two algorithms were presented for road classification and road profile observation. First, transfer function-based algorithms were introduced. A simple transfer function model was derived, and then a two-step road classifier was presented. The unknown road excitation was attributed to a class according to the PSD at the central frequencies, and a road level determination procedure was then provided using fitted waviness. Simulation validation was performed for examining the effects of varying velocity, different inputs for the RF classifier, noisy measurements, and complex tire modeling. A field test was finally performed to validate the algorithm, and more than 75% classification accuracy was observed with good robustness to changing conditions. An adaptive unknown road observer was then introduced to estimate road profiles in the time domain. The observer was composed of two individual observers, i.e., AKF and ASTO. The unsprung mass velocity observed by the AKF is treated as the input of ASTO, by which the unknown road profile was estimated. The noise covariance of the AKF was tuned using a road classification algorithm to improve estimation accuracy. Finally, a simulation validation was performed for road level ISO-A, B, and C. Simulation results indicated that the proposed AKF-ASTO had higher accuracy compared to a traditional method with measurements of only sprung mass and unsprung mass accelerations.

CHAPTER 5

Road Adaptive Hybrid Suspension Control

In Chapters 5 and 6, we will present two examples applying the introduced road estimation algorithms. In this chapter, an adaptive hybrid suspension controller is formulated based on the road classification algorithm introduced in Chapter 3. This chapter first provides analytical expressions of a quarter vehicle suspension system for sprung mass acceleration (SMA), rattle space (RS), and tire deflection (TD), and then details the influence of system parameters and road levels on the responses. Suspension control is then transformed into a MOOP, which is resolved using NSGA-II. Based on the above analysis, a road adaptive hybrid suspension controller is finally proposed to improve vehicle vertical performance.

5.1 CORRELATION BETWEEN SUSPENSION PARAMETER AND ROAD EXCITATION

This section first presents analytical expressions for vehicle sprung mass acceleration, rattle space, and tire deflection with respect to road input and vehicle velocity. Next, the influence of suspension parameters and road conditions on the responses is investigated.

5.1.1 ANALYTICAL EXPRESSIONS OF VEHICLE RESPONSES

Considering passive suspension dynamics depicted by Eq. (2.39), the transfer functions of SMA, RS, and TD with respect to road profile can be calculated as follows:

$$H(s)_{\ddot{x}_b \sim x_r} = \frac{s^2 X_b(s)}{X_r(s)} = \frac{c_p k_t s^3 + k_s k_t s^2}{A}, \tag{5.1}$$

$$H(s)_{(x_w - x_r) \sim x_r} = k_t \left(\frac{X_w(s)}{X_r(s)} - 1 \right) = k_t \frac{D}{A}, \tag{5.2}$$

$$H(s)_{(x_b - x_w) \sim x_r} = \frac{X_w(s)}{X_r(s)} = \frac{-m_b k_t s^2}{A}, \tag{5.3}$$

where

$$A = m_w m_b s^4 + \left(c_p m_b + c_p m_w \right) s^3 + \left(m_w k_s + m_b k_s + m_b k_t \right) s^2 + c_p k_t s + k_t k_s$$

$$D = -m_b m_w s^4 - \left(c_p m_b + c_p m_w \right) s^3 - \left(m_b k_s + m_w k_s \right) s^2.$$

For the suspension system excited by a road profile, which can be considered as a stationary process, the variance of the above responses can be expressed as [111]:

$$\sigma_p^2 = \frac{1}{2\pi} \int_{-\infty}^{\infty} S_p(\omega) \, d\omega, \tag{5.4}$$

where $S_p(\omega)$ represents the PSD of response p, which is expressed as:

$$S_p(\omega) = \left| H(j\omega)_{p\sim x_r} \right|^2 S_{x_r}(j\omega), \tag{5.5}$$

where $S_{x_r}(j\omega)$ is the road PSD, and $H(j\omega)_{p\sim x_r}$ is the frequency response function from road excitation to system response p. Newton et al. proposed an analytical expression for a response whose PSD can be expressed by the following form [112]:

$$S_p(\omega) = \frac{N_{k-1}(j\omega)N_{k-1}(-j\omega)}{D_k(j\omega)D_k(-j\omega)}, \tag{5.6}$$

where $D_k(\cdot)$ and $N_{k-1}(\cdot)$ represent two polynomials with the order of kth and $(k-1)^{th}$, which can be expressed as:

$$D_k(j\omega) = d_k(j\omega)^k + d_{k-1}(j\omega)^{k-1} + \cdots + d_0 \tag{5.7}$$

$$N_{k-1}(j\omega) = n_{k-1}(j\omega)^{k-1} + n_{k-2}(j\omega)^{k-2} + \cdots + n_0. \tag{5.8}$$

In the following, discussions of the decomposition of $H(j\omega)_{p\sim x_r}$ and $S_{x_r}(j\omega)$ are given, and the road profile is generated using integrated white noise and rational function algorithms, which were discussed in Section 2.1.4.

CASE 1. System responses based on the integrated white noise algorithm.   According to Eqs. (2.16)–(2.20), road profile PSD generated by the integrated white noise algorithm can be expressed as:

$$G_q(f) = \frac{A_{xr}v}{f^2} \rightarrow S_{x_r-1}(j\omega) = A_{xr}v\frac{1}{j\omega}\frac{1}{-j\omega}, \tag{5.9}$$

where $A_{xr} = 4\pi^2 n_0^2 G_{x_r}(n_0)\,v$.

Since $\left| H(j\omega)_{p\sim x_r} \right|^2 = H(j\omega)_{p\sim x_r} H(-j\omega)_{p\sim x_r}$, and for $S_{x_r-1}(j\omega)$, it can be shown that the polynomial order is $k = 4$. Therefore, the response variance, σ_p^2, can be given as [111]:

$$\sigma_{p-1}^2 = \frac{\left[n_3^2 c_m + (n_2^2 - 2n_1 n_3)d_0 d_1 d_4 + (n_1^2 - 2n_0 n_2)d_0 d_3 d_4 + n_0^2\left(-d_1 d_4^2 + d_2 d_3 d_4\right)\right]}{2d_0 d_4(-d_0 d_3^2 - d_1^2 d_4 + d_1 d_2 d_3)}, \tag{5.10}$$

where $c_m = -d_0^2 d_3 + d_0 d_1 d_2$.

The analytical expressions of SMA, RS, and TD are then expressed as follows.

1. SMA

$$\sigma^2_{\ddot{x}_b-1} = \frac{A_{x_r}v}{2m_b^2} \cdot \left[\frac{(m_b+m_w)\,k_s^2}{c_p} + k_t c_p\right].$$ (5.11)

2. RS

$$\sigma^2_{(x_b-x_w)-1} = \frac{A_{x_r}v}{2} \cdot \frac{m_b+m_w}{c_p}.$$ (5.12)

3. TD

$$\sigma^2_{(x_w-x_r)-1} = \frac{A_{x_r}v}{2} \cdot (m_b+m_w)^2$$
$$\left[\frac{(m_b+m_w)\,k_s^2}{m_b^2 c_p} - \frac{2k_s k_t m_w}{m_b c_p\,(m_b+m_w)} + \frac{k_t^2 m_w}{c_p\,(m_b+m_w)^2} + \frac{k_t c_p}{m_b^2}\right].$$ (5.13)

CASE 2. System responses base on the rational function algorithm. According to Eq. (2.13), the road profile PSD generated by the rational function algorithm can be expressed as:

$$G_q(f) = \frac{av\rho^2}{\pi\left[(av)^2 + f^2\right]} \rightarrow S_{x_r-2}(j\omega) = \frac{av\rho^2}{\pi} \frac{1}{(av+j\omega)} \frac{1}{(av-j\omega)}.$$ (5.14)

Since $\left|H\,(j\omega)_{p\sim x_r}\right|^2 = H\,(j\omega)_{p\sim x_r}\,H\,(-j\omega)_{p\sim x_r}$, and for $S_{x_r-2}\,(j\omega)$, it can be calculated that the polynomial order is $k = 5$. The response variance, σ_p^2, can be given as follows [111]:

$$\sigma_p^2 = \frac{(n_4^2 c_{m0} + (n_3^2 - 2n_2 n_4)c_{m1} + n_{m0}c_{m2} + (n_1^2 - 2n_0 n_2)c_{m3} + n_0^2 c_{m4})}{2d_0(d_1 c_{m4} - d_3 c_{m3} + d_5 c_{m2})},$$ (5.15)

where $n_0,\ldots,n_4$ and $d_0,\ldots,d_5$ are coefficients defined in Eq. (5.10), and $n_{m0},c_{m0},\ldots,c_{m4}$ are defined as:

$$n_{m0} = n_2^2 - 2n_1 n_3 + 2n_0 n_4, \quad c_{m0} = \frac{d_3 c_{m1} - d_1 c_{m2}}{d_5}, \quad c_{m1} = -d_0 d_3 + d_1 d_2;$$

$$c_{m2} = -d_0 d_5 + d_1 d_4, \quad c_{m3} = \frac{d_2 c_{m2} - d_4 c_{m1}}{d_0}, \quad c_{m4} = \frac{d_2 c_{m3} - d_4 c_{m2}}{d_0}.$$

The analytical expressions of SMA, RS, and TD can then be expressed as follows.

1. SMA

$$\sigma^2_{\ddot{x}_b-2} = \frac{av\rho^2}{2\pi}$$
$$\cdot \frac{k_t^2 c_p^2 \left(k_s + c_p av + m_b a^2 v^2\right) + k_t k_s^2 \left[(m_b + m_w)\left(k_s + c_p av\right) + m_b m_w a^2 v^2\right]}{m_b^2 c_p M}.$$

(5.16)

2. RS

$$\sigma^2_{(x_b - x_w)-2} = \frac{av\rho^2}{2\pi} \cdot \frac{k_t \left(k_s m_w + k_s m_b + m_w c_p av + m_b m_w a^2 v^2\right)}{c_p M}.$$

(5.17)

3. TD

$$\sigma^2_{f_D-2} = \frac{av\rho^2}{2\pi} \cdot \frac{P + Q + R}{k_t^2 m_b^2 c_p^2 M},$$

(5.18)

where:

$$P = k_t^4 \left(m_b + m_w\right) c_p \left(-2k_s m_b m_w + m_w c_p^2 + m_b c_p^2\right)\left(k_s + c_p av + m_b a^2 v^2\right)$$
$$Q = k_t^3 k_s^2 \left(m_b + m_w\right)^2 c_p \left(k_s m_w + k_s m_b + m_w c_p av + m_b c_p av + m_b m_w a^2 v^2\right)$$
$$R = k_t^4 m_w m_b^2 c_p \left[k_s k_t + k_t c_p av + (k_s + k_t) m_b a^2 v^2 + m_b c_p a^3 v^3\right].$$

From Eqs. (5.11)–(5.13), it can be seen that the system responses are in direct proportion to both road level and vehicle velocity for the integrated white noise algorithm. This means that a designer must focus on the suspension parameters instead of road and speed information to improve vehicle performance. As for the rational function algorithm, the coupling effect between parameters and excitation conditions is unavoidable. In this case, how to adjust suspension parameters for varying road levels and vehicle speeds becomes a key issue. The next section provides an in-depth analysis of the relationship between suspension systems and excitation conditions for the two road generation algorithms.

5.1.2 CORRELATION BETWEEN SUSPENSION SYSTEM AND EXCITATION CONDITIONS

This section provides more details on the correlation between suspension system and excitation conditions. For the suspension system, both suspension stiffness k_s and damping coefficient c_p are regarded as the key variables, and system responses are analytically analyzed with varying vehicle velocity v. All the suspension parameters along with velocity are given in Table 5.1.

Table 5.1: System parameters

Parameter	Value	Variation
m_b (kg)	410	N/A
m_w (kg)	39	N/A
k_t (N/m)	183,000	N/A
c_p (Ns/m)	2,000	600-2,800 (100 Ns/m interval)
k_s (N/m)	20,000	10,000-40,000 (1000 N/m interval)
ISO level	D	N/A
v (m/s)	20	5-40 (5 m/s interval)

In Table 5.1, the variation range of suspension stiffness k_s corresponds to 0.78–1.57 Hz for the sprung mass natural frequency. The variation of the damping coefficient c_p results in a damping ratio range from 0.1–0.49. The velocity, v, variation covers normal conditions ranging from 18–144 km/h. Note that the main purpose of this section is to provide an in-depth analysis of the correlation between suspension system and excitation conditions. The variation shown in Table 5.1 covers normal suspension models in practice [113]. In the following, the red marker "*" on the X–Y plain represents the minimal response value on the Z axis under the variation of the variable shown on the X axis.

CASE 1. System responses based on the integrated white noise algorithm.   Here, we introduce the influence of system parameters and vehicle speed to system responses for roads generated by the integrated white noise algorithm.

1. SMA

   The standard derivations (STD) of SMA for varying parameters and velocity are shown in Figure 5.1. The following can be concluded.

   (a) The STD of SMA is in direct proportion to vehicle velocity, and larger vehicle speeds will cause increasingly sever vehicle SMA.

   (b) Figure 5.1a and b reveal that the minimal value of SMA is insensitive to vehicle velocity.

   (c) Figure 5.1a and b show that the SMA can be reduced by changing the suspension stiffness and the damping. The minimal values can be obtained by setting $k_s = 10,000$ N/m and $c_p = 1,000$ Ns/m.

   (d) Figure 5.1c indicates that the SMA is determined by the suspension stiffness and damping when velocity is fixed. Increasing suspension stiffness requires larger damping to obtain the minimal SMA.

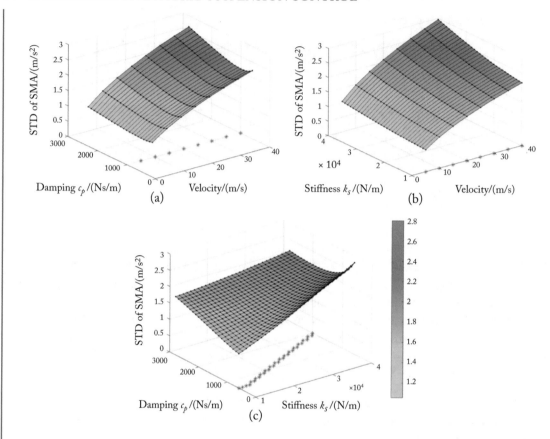

Figure 5.1: SMA for road generated by integrated white noise algorithm: (a) influence of damping and vehicle velocity, (b) influence of stiffness and vehicle velocity, and (c) influence of stiffness and damping.

2. RS

   (a) The STD of RS is in direct proportion to vehicle velocity, and larger vehicle speeds cause larger RS amplitudes.

   (b) Figure 5.2a reveals that the minimal value of RS is insensitive to vehicle velocity.

   (c) Figure 5.2a shows that the optimal damping for the investigated system is 2800 Ns/m, and the RS is insensitive to suspension stiffness for the road generated by the integrated white noise algorithm. This can also be observed by examining Eq. (5.12).

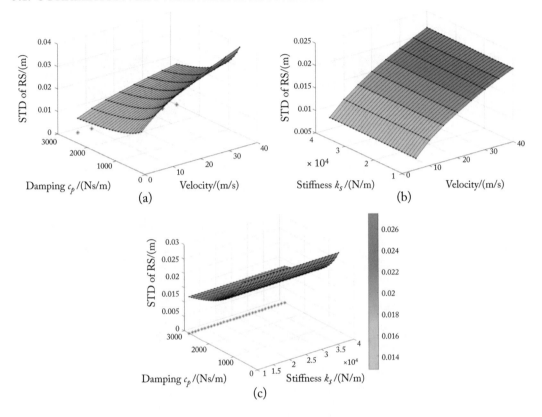

Figure 5.2: RS for road generated by integrated white noise algorithm: (a) influence of damping and vehicle velocity, (b) influence of stiffness and vehicle velocity, and (c) influence of stiffness and damping.

3. TD

   (a) The STD of TD is in direct proportion to vehicle velocity.

   (b) Figure 5.3a and b reveal that the minimal value of TD is insensitive to vehicle velocity.

   (c) Figure 5.3a and b show that the minimal values of TD can be obtained by setting $k_s = 15{,}000$ N/m and $c_p = 2300$ Ns/m.

   (d) Figure 5.3c indicates that the TD is determined primarily by suspension stiffness and damping when the vehicle velocity is fixed. Increasing suspension stiffness demands larger damping to obtain the minimal TD.

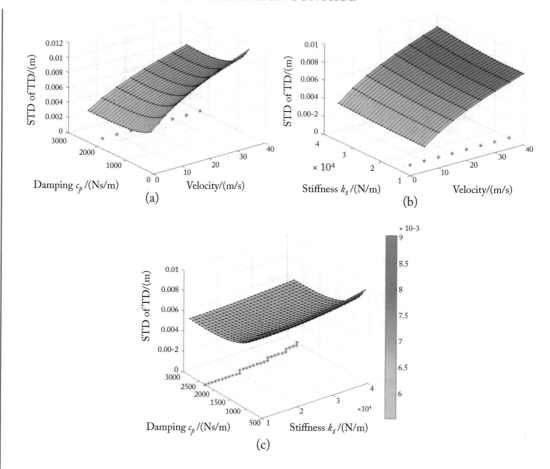

Figure 5.3: TD for road generated by integrated white noise algorithm: (a) influence of damping and vehicle velocity, (b) influence of stiffness and vehicle velocity, and (c) influence of stiffness and damping.

CASE 2. System responses base on the rational function algorithm.

1. SMA

The STD of SMA for varying parameters and vehicle velocity are shown in Figure 5.4. The following observations can be made regarding the rational function algorithm.

(a) The STD of SMA is in direct proportion to vehicle velocity, and larger vehicle speeds will cause more severe vehicle SMA.

(b) Figure 5.4a indicates that the increasing velocity causes the damping coefficient to reduce from 900 Ns/m to 800 Ns/m to obtain the minimal SMA. As for Figure 5.4b, we can conclude that the stiffness has little effect on SMA for increased vehicle speeds.

(c) Figure 5.4c shows that the SMA is determined by both suspension stiffness and damping. As well, increasing suspension stiffness requires larger damping to obtain the minimal SMA.

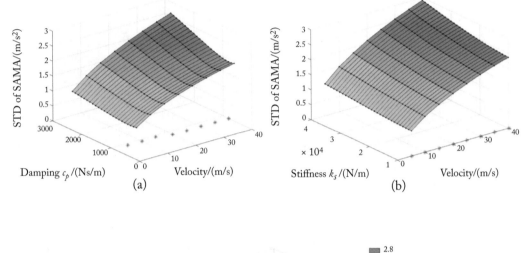

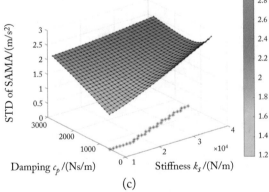

Figure 5.4: SMA for road generated by rational function algorithm: (a) influence of damping and vehicle velocity, (b) influence of stiffness and vehicle velocity, and (c) influence of stiffness and damping.

2. RS

    (a) The STD of RS is in direct proportion to vehicle velocity, and larger vehicle speed will cause more severe vehicle RS.

    (b) Figure 5.5a and b indicate that RS is insensitive to both stiffness and damping coefficient. As well, the minimal RS is independent of vehicle velocity. This can also be observed from the analytical expression shown in Eq. (5.17).

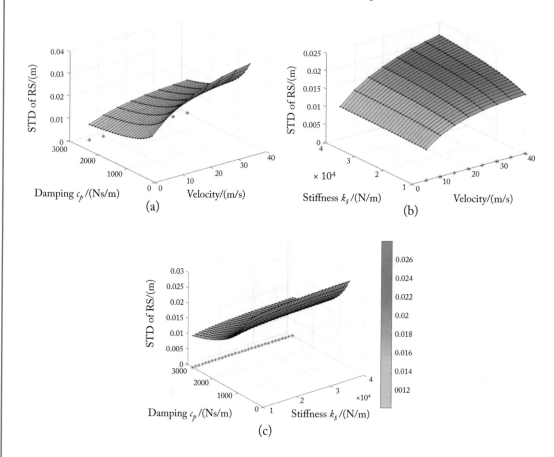

Figure 5.5: RS for road generated by rational function algorithm: (a) influence of damping and vehicle velocity, (b) influence of stiffness and vehicle velocity, and (c) influence of stiffness and damping.

3. TD

    (a) The STD of SMA is in direct proportion to vehicle velocity, and larger vehicle speed will cause more severe vehicle SMA.

(b) Figure 5.6a indicates that the increasing velocity causes the damping coefficient to reduce from 900 Ns/m to 800 Ns/m to obtain the minimal SMA. As for Figure 5.6b, we can conclude that the stiffness has little effect on SMA for the increased vehicle speed.

(c) Figure 5.6c shows that the SMA is determined by both suspension stiffness and damping. As well, increasing suspension stiffness requires larger damping to obtain the minimal SMA.

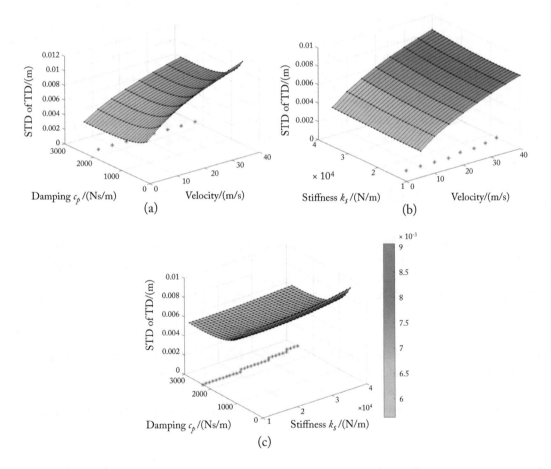

Figure 5.6: TD for road generated by rational function algorithm: (a) influence of damping and vehicle velocity, (b) influence of stiffness and vehicle velocity, and (c) influence of stiffness and damping.

The following conclusions can be made based on the analysis for the two road profile algorithms.

(a) For a suspension system excited by the two generated road profiles, smaller damper and stiffness can result in better ride comfort, whereas increased damping ensures better handling ability. Since the natural frequency of the sprung mass, which is typically 1–1.5 Hz for passenger vehicles, restricts the range of suspension stiffness, the suspension damper coefficient dominates vehicle vertical performance and plays an important role in the competitive balance between ride comfort and road handling.

(b) Variation of road conditions affects vehicle vertical dynamics, and the influences are twofold. First, high vehicle velocity represents increased road excitation energy. This results in worse ride comfort, road handling, and rattle space. For the rational function algorithm, vehicle velocity will influence the optimal suspension parameters. Conversely, vehicle speed is irrelevant to that of the integrated white noise algorithm. This conclusion can be observed from Eqs. (5.11)–(5.13) and Eqs. (5.16)–(5.18).

Due to the contradiction between ride comfort and road handling, selection of suspension stiffness and damping coefficient remains a challenge for passive suspension systems. The following section will illustrate how to design a road adaptive suspension controller based on the above analysis.

5.2 MULTI-OBJECTIVE OPTIMIZATION PROBLEM AND SOLUTION

MOOP is a common issue in academia and the industry. Single objective optimization problems are a particular case of MOOP when only one objective is to be optimized [114]. This part here introduces the concepts of MOOP, and then uses a non-dominated sorting genetic algorithm-II (NSGA-II) to solve the MOOP for the suspension system examined here.

5.2.1 MOOP

The core of MOOP is to find a set of solutions that can balance contradictory or competitive objective functions. Generally, a solution in the set cannot optimize all of the objective functions simultaneously, and the concept of non-dominated set is defined to solve this issue (where none of the solutions can be said to be better than any others with respect to all of the objectives) [115]. A MOOP in its general form can be stated as follows:

$$
\begin{cases}
\min \ y = F(x) = f_m(x), & m = 1, 2, \ldots, M \\
\text{s.t. } g_j(x) \geq 0, & j = 1, 2, \ldots, J \\
h_k(x) = 0, & k = 1, 2, \ldots, K \\
x_i^{(L)} \leq x_i \leq x_i^{(U)}, & i = 1, 2, \ldots, N
\end{cases}
\tag{5.19}
$$

where $x = (x_1, \ldots, x_N) \in X \subset \Re^N$ is a vector of N decision variables, and the constraints of the MOOP are J inequalities, K equalities, and N bounds. All solutions that satisfy all of the constraints and bounds compose the feasible region of the problem. $y = f_m(x) \in Y \subset \Re^M$ donates the objective function set, and y is the objective space. The mapping function, $F(x)$, transforms solution x to a point in the objective space. Several definitions that will be used in the following sections are given below.

Definition 5.1   Feasible solution set X.   For any $x \in \mathbf{X}$, if $g_j(x) \geq 0$, $h_k(x) = 0$, and $x_i^{(L)} \leq x_i \leq x_i^{(U)}$, then x is a feasible solution in the feasible solution set **X**.

Definition 5.2   Pareto dominance.   For any $x_A, x_B \in \mathbf{X}$, if

$$\forall p = 1, 2, \ldots, M, \quad \forall q = 1, 2, \ldots, M, \quad f_p(x_A) < f_q(x_B), \quad p \neq q \quad (5.20)$$

hold, then x_A dominates x_B, which can be expressed as $x_A \succ x_B$.

Definition 5.3   Pareto optimal set.   All Pareto optimal solution x^* compose the Pareto optimal set P^*, which is defined as:

$$P^* = \{x^* | \forall x \in \mathbf{X}, \mathbf{x}^* \succ x\}. \quad (5.21)$$

Definition 5.4   Pareto front set.   Pareto front set PF^* includes all the objective function values mapping from the solutions in the Pareto optimal set P^*.

$$PF^* = \{F(x^*) | \mathbf{x}^* \in P^*\}. \quad (5.22)$$

Definition 5.5   Fitness function.   A fitness function is used to evaluate the dominance of the solutions, and is typically the same as the objective function.

The relationship of the above definitions can be graphically shown in Figure 5.7. The shadow area in the left subplot represents the feasible region, and the shadow in the right subplot is the corresponding objective function value. The dash lines between the two subplots illustrates the mapping $F(x)$.

5.2.2   SOLUTION OF MOOP

Finding a solution to a MOOP is a hot topic and has attracted significant attention in recent decades [115, 116]. Classical solvers typically solve the MOOP by transforming a MOOP to several single objectives. Some typical examples of solvers include the weighted sum method, ε-constraint method, weighted metric method, Benson's method, and goal programming method. By analyzing these methods, Deb pointed out the following difficulties when multiple Pareto optimal solutions are desired [115].

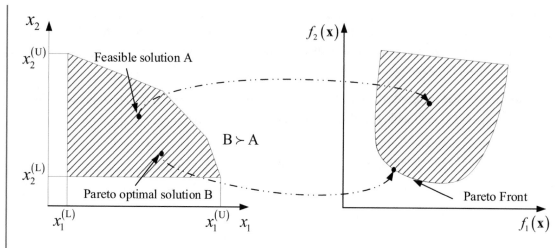

Figure 5.7: MOOP and Pareto optimal solution.

1. Classical methods can provide only one optimal solution in one simulation run.

2. The methods cannot guarantee the completeness of the Pareto optimal solution set for nonconvex MOOPs.

3. Prior knowledge of the problem being studied is required by all the mentioned methods.

Apart from the aforementioned issues, complex MOOPs in academia and the industry contain intricate nonlinear dynamics, and the efficiency of the classical methods cannot meet many requirements. Researchers have begun to apply natural evolutionary principles to find the Pareto front solution set. Schaffer et al. proposed a vector evaluated algorithm, which was the first method of solving a MOOP based on genetic algorithm (GA) [117]. Deb et al. [118], Fonseca et al. [119], and Horn et al. [120] then presented NSGA, multi-objective GA, and Niche GA, separately. These algorithms compose the first generation of these techniques, and common features are non-elitist and a reliance of sharing parameters. Second-generation methods were then proposed to remedy the issues that appeared in the previous generation (i.e., high complexity of sorting, lack of elitism, and a need for sharing parameters) [121]. Some representative second-generation algorithms are NSGA-II [121], strength Pareto approach [122], and envelope-based selection algorithm [123].

In this section, NSGA-II is used to solve the MOOP of the suspension system. A flowchart of this algorithm is shown in Figure 5.8 and Table 5.2.

In the next part, the MOOP of the suspension system will be solved using NSGA-II.

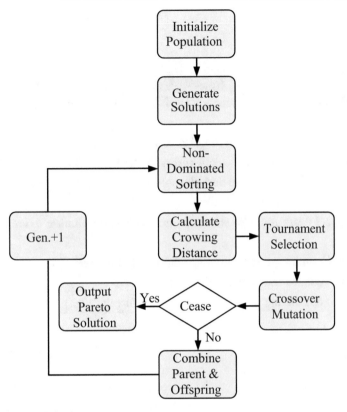

Figure 5.8: Flowchart of NSGA-II.

5.3   ROAD ADAPTIVE HYBRID SUSPENSION CONTROLLER

In this part, a road adaptive hybrid controller is formulated to improve suspension performance for varying road conditions.

5.3.1   HYBRID CONTROL AND ITS ANALYTICAL EXPRESSIONS

Hybrid control algorithms combine the advantages of skyhook and groundhook control by applying a weighted sum force, F_d, to the suspension system. The models for both the ideal hybrid suspension system and its equivalent system are shown in Figure 5.9:

$$F_d = c_{sky} \cdot \dot{x}_b - c_{grd} \cdot \dot{x}_w. \tag{5.23}$$

In the left subplot of Figure 5.9, two ideal dampers connect the sprung and the unsprung masses to an inertial reference. As this structure is unrealistic, the equivalent model on the right uses a controllable damping force F_d to mitigate the vibration induced by the road disturbance.

Table 5.2: NSGA-II procedure

Procedure of NSGA-II
Step 1. **Initialization**. Select the parameters for NSGA-II, e.g., elites number and population size.
Step 2. **Generate solutions**. Calculate initial random solutions within the feasible region.
Step 3. **Non-dominated sorting**. Calculate and rank the fitness function to form the fronts.
Step 4. **Calculate crowding distance**. Sort the solutions in descending order and select the solutions with higher crowding distance.
Step 5. **Selection and mutation**. Use selection, crossover, and mutation to create a new population.
Step 6. **Combine**. Combine parent and offspring populations and choose the best solutions with a size equal to the predefined population size.
Step 7. **Cease**. Stop optimizing until the fitness function values reach the desired threshold.

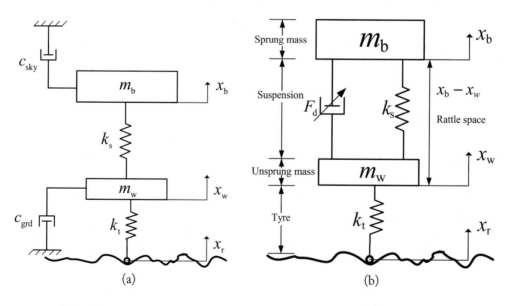

Figure 5.9: Hybrid suspension system: (a) ideal model and (b) equivalent model.

The dynamic equations of Figure 5.9b are given below:

$$m_b \ddot{x}_b + k_s (x_b - x_w) + F_d = 0$$
$$m_w \ddot{x}_w + k_s (x_w - x_b) + k_t (x_w - x_r) - F_d = 0. \tag{5.24}$$

By taking Laplace transform for Eq. (5.24), the relationships of sprung mass acceleration, tire deflection, and rattle space with respect to road profile are calculating as follows:

$$H(s)_{hybrid-\ddot{x}_b \sim x_r} = \frac{c_{grd} k_t s^3 + k_s k_t s^2}{A}, \tag{5.25}$$

$$H(s)_{hybrid-f_D \sim x_r} = \frac{D}{A}, \tag{5.26}$$

$$H(s)_{hybrid-(x_b - x_w) \sim x_r} = \frac{-m_b k_t s^2 + (c_{grd} k_t - c_{sky} k_t) s}{A}, \tag{5.27}$$

where

$$A = m_w m_b s^4 + (c_{grd} m_b + c_{sky} m_w) s^3 + (m_w k_s + m_b k_s + m_b k_t) s^2 + c_{sky} k_t s + k_t k_s$$
$$D = -m_b m_w s^4 - (c_{grd} m_b + c_{sky} m_w) s^3 - (m_b k_s + m_w k_s) s^2.$$

Similar to the procedure introduced in Section 5.1, the analytical expressions of sprung mass acceleration, tire deflection, and rattle space with respect to road conditions can be derived. The variables of these functions include vehicle velocity, road level, c_{sky}, and c_{grd}. In the following part, the vehicle velocity is assumed to be a constant, and only controller parameters c_{sky} and c_{grd} are to be tuned online if the road classification algorithms in Chapter 3 are adopted. The hybrid suspension control problem is then be transformed into a MOOP. For more details on the analytical expressions, refer to Qin et al. [124].

5.3.2   ROAD ADAPTIVE HYBRID CONTROLLER

The hybrid suspension control system is first transformed into the following MOOP:

$$\begin{aligned}
\min \quad & f_1(c_{sky}, c_{grd}) = \sigma_{\ddot{x}_b}^2; \\
& f_2(c_{sky}, c_{grd}) = \sigma_{TD}^2, \\
\text{s.t.} \quad & 6 \cdot |\sigma_{x_b - x_w}| \le \lim(x_b - x_w), \\
& 500 \, (\text{Ns/m}) \le c_{sky} \le 5000 \, (\text{Ns/m}), \\
& 500 \, (\text{Ns/m}) \le c_{grd} \le 5000 \, (\text{Ns/m}),
\end{aligned} \tag{5.28}$$

where $\sigma_{\ddot{x}_b}$, σ_{TD}, and $\sigma_{x_b - x_w}$ are the standard deviation of sprung mass acceleration, tire deflection, and rattle space, respectively. The limit $\lim(x_b - x_w)$ represents the rattle space limitation, which is selected as ± 60 mm. The lower and upper bounds of both c_{sky} and c_{grd} are chosen as 500 Ns/m and 5000 Ns/m.

Now, the presented MOOP in Eq. (5.28) will be solved to illustrate the relationship between the Pareto optimal solution set and suspension parameters. The following parameters are adopted in this analysis: the solution number of the Pareto front is 50, the population size is 100, and the maximum generation is 200. The result of this analysis is shown in Figure 5.10.

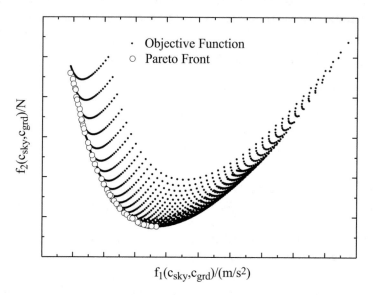

Figure 5.10: Pareto front and objective function.

In Figure 5.10, the dots represent the objective function values calculated by varying c_{sky} and c_{grd} within the bounds, and the circular dots are the Pareto Front. We can observe that the Pareto optimal solutions are located at the bottom left of the objective function area, and none of the optimal solutions dominates the others. This means that there is no solution in the Pareto Front that outperforms other solutions in both aspects of ride comfort and road handling.

We can further observe from Figure 5.10 that the investigated MOOP is a convex optimization problem. Therefore, the contradiction between ride comfort and road handling can be resolved by varying c_{sky} and c_{grd}. The Pareto Front obtained by NSGA-II contains several solutions that can be used for different road conditions, and the choice of the weight can be subjective. One widely adopted principle that will be used in this part is stated as follows [28, 99, 125, 126]:

1. road handling-oriented mode and

2. ride comfort-oriented mode.

Since worse road condition increases input energy, which increases dynamic variations of tire deflection, we assign greater weighting to tire deflection to reduce σ_{TD}. In contrast, a greater weighting for ride comfort is desired for good road conditions to enhance passenger experience.

Based on the above principles, the damping coefficients for different road levels are calculated offline and tabulated in Table 5.3.

Table 5.3: Controller weightings and parameters

Road Level	Weighting $[w_{ACC}, w_{TD}]$	Parameter $[c_{sky}, c_{grd}]$
Good (ISO-A, B)	[0.8, 0.2]	[4567, 651]
Average (ISO-C, D)	[0.6, 0.4]	[4046, 960]
Poor (ISO-E, F)	[0.4, 0.6]	[3234, 1425]
Very poor (ISO-G, H)	[0.2, 0.8]	[2750, 1680]

Table 5.3 reveals that by increasing w_{TD} for poor road conditions, the resulting increased c_{grd} and decreased c_{sky} can provide better handling capacity with larger F_{grd}. With the calculated damping coefficients, the structure of the road adaptive hybrid suspension controller is given in Figure 5.11.

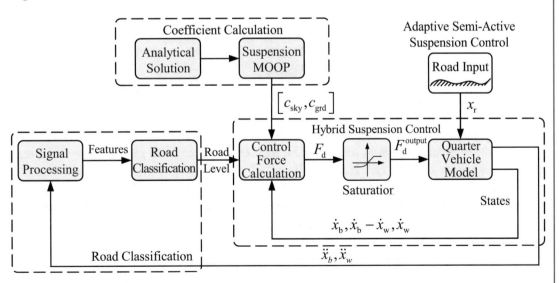

Figure 5.11: Road adaptive hybrid suspension controller.

It can be seen from Figure 5.11 that the proposed algorithm is composed of three parts: coefficient calculation, hybrid suspension control, and road classification. The coefficient calculation outputs the coefficients $[c_{sky}, c_{grd}]$ by solving the MOOP in Eq. (5.28), and the results are then stored for online application. The hybrid suspension control then calculates the ideal force as per the damping coefficients and system states. Finally, the actual control force is applied in the suspension system after saturation. The system acceleration responses are then used for road classification, which can provide accurate road level estimation and close the feedback loop.

The road excitation with road level ISO C-F is adopted to evaluate performance of the proposed controller.

We can observe from Table 5.4 that poor road conditions result in severer system responses for both the passive and the proposed systems. By changing road level from ISO B to ISO F, the sprung mass accelerations of the passive and the proposed systems increase from 0.371 m/s^2 to 5.806 m/s^2, and 0.311 m/s^2 to 6.617 m/s^2, respectively. Additionally, the tire deflection of the passive and the proposed systems vary from 0.00117–0.0188 m and 0.00125–0.0179 m, respectively. This shows that the proposed algorithm can provide better ride comfort for good roads and improved handling for bad roads. Comparisons of the results in the frequency domain are presented in Figures 5.12 and 5.13.

Table 5.4: Comparison of ride comfort and handling capacity for different conditions

Road Level (ISO)	RMS of SMA (m/s^2)			RMS of TD (m)		
	Passive	Proposed	Improvement	Passive	Proposed	Improvement
B	0.371	0.311	16.2%	0.00117	0.00125	-6.88%
D	1.496	1.435	4.1%	0.00473	0.00474	-0.14%
E	2.921	3.305	-13.1%	0.00935	0.00888	4.97%
F	5.806	6.617	-13.9%	0.0188	0.0179	5.01%
C	0.752	0.706	6.1%	0.00228	0.00228	-0.23%

In Figure 5.12, it is seen that by increasing w_{ACC}, the amplitude of SMA is reduced at sprung mass resonant frequency, within the human sensitive frequency range, i.e., 4–8 Hz, and frequencies higher than 10 Hz. It should be noted that all four weightings can mitigate the vibration at the sprung mass resonant frequency. In Figure 5.13, it can be observed that increased w_{tire} results in the reduction of vibration in the unsprung mass resonant frequency. However, the controller performance between 4–8 Hz is deteriorated when w_{tire} is increased.

Next, the influence of vehicle velocity on the controller gains is investigated. Here, we use NSGA-II method to solve the MOOP with different vehicle velocities. Simulation results obtained for vehicles velocities ranging from 20–120 km/h and road level ISO-B, D, F, and H are presented in Figure 5.14 and Table 5.5.

In Figure 5.14, the bar heights represent the damping coefficients at 40 km/h. The error bars are determined by the extreme damping values at different velocities. The bars in magenta represent c_{sky}, and the bars in black represent c_{grd}. It can be seen from Figure 5.14 and Table 5.5 that the variation of controller gains relatively insignificant for varying velocities. Therefore,

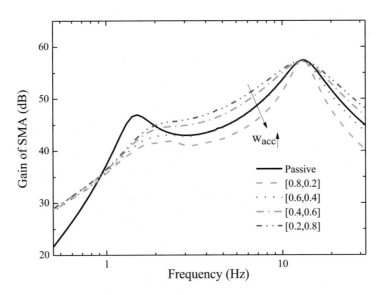

Figure 5.12: Comparison of frequency responses of $\ddot{x}_b/x_r$ for different control weights.

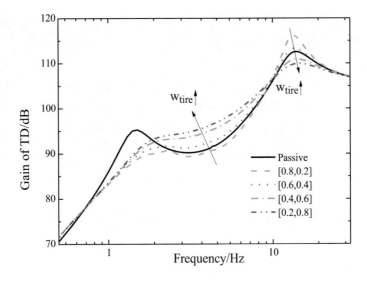

Figure 5.13: Comparison of frequency responses of f_D/x_r for different control weights.

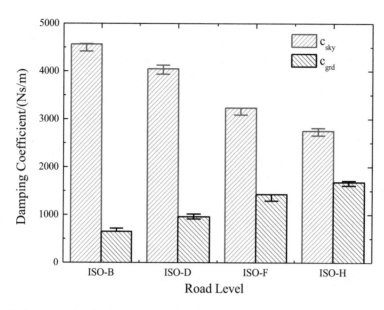

Figure 5.14: Influence of vehicle velocity to the controller.

Table 5.5: Influence of vehicle velocity to the damping coefficient

Level Velocity (km/h)	ISO-B $[c_{sky}, c_{grd}]$	ISO-D $[c_{sky}, c_{grd}]$	ISO-F $[c_{sky}, c_{grd}]$	ISO-H $[c_{sky}, c_{grd}]$
20	[4451,652]	[4122,912]	[3134,1356]	[2684,1646]
40	[4567,651]	[4046,960]	[3234,1425]	[2750,1680]
60	[4415,718]	[3934,972]	[3178,1407]	[2712,1721]
80	[4550,703]	[4044,1017]	[3089,1369]	[2654,1655]
100	[4575,658]	[4087,981]	[3210,1290]	[2782,1611]
120	[4437,649]	[4123,956]	[3198,1387]	[2811,1712]

we can apply constant damping coefficients (for instance the value at 40 km/h) in real-world applications.

5.4   SUMMARY

This chapter presented a road adaptive semi-active suspension control algorithm to improve vehicle ride comfort and handling capacity. To this end, we first provided the analytical expressions of vehicle sprung mass acceleration, rattle space, and tire deflection with respect to road input and vehicle velocity. Next, the influence of suspension parameters and road generation algorithms on the vehicle response was investigated. The concepts of MOOP were then introduced. The MOOP for the present suspension system was formulated and resolved using NSGA-II. A road adaptive hybrid controller was finally proposed to improve suspension performance for varying road conditions. Simulation results demonstrated that the proposed algorithm could effectively reduce suspension system vibrations and different suspension modes could be utilized by changing the controller weighting.

CHAPTER 6

Suspension Predictive Control Based on Road Estimation

This chapter presents predictive semi-active suspension controllers. A hybrid model predictive controller (HMPC) is presented to improve system performance by considering the constraints of the controllable damper and system states. Different from previous research, road profiles in the time domain are estimated and compensated using the HMPC. The structure of this chapter is as follows. First, the concepts of model predictive and optimal predictive control are introduced. Next, an HMPC is presented to compensate for road disturbances and improve system performance. A simulation study is finally performed for both quarter and half vehicle models.

6.1 HYBRID MODEL PREDICTIVE CONTROL

In this section, we introduce model predictive control, and then present the concepts of hybrid system and hybrid model predictive control.

6.1.1 MODEL PREDICTIVE CONTROL

Optimal control for systems with constraints has been a hot topic in the study of control engineering for decades. Two main issues restrict the application of the traditional algorithms:

1. the constraints in a linear system may result in nonlinear feedback law [127]; and

2. optimal control for systems with constraints inherently involve solving quadratic programming or linear programming problems for systems with infinite states and constraints which are also related to initial conditions [128].

In this case, solving a receding horizon optimization problem becomes a possible solution for optimal control of a system with constraints [129–132]. To this end, model predictive control (MPC) was proposed in the 1980s and has been widely applied in the chemical industry, process control in petroleum refineries [133], and many other fields [134, 135]. The biggest advantage of MPC is that it can predict system states and provide the optimal solution for a system with states, control, and output constraints. These merits have inspired researchers to apply MPC in suspension control. Cho et al., Mehra et al., Donahue et al., and Gopalasamy et al. studied the application of MPC in active suspension systems. In these works, rattle space was regarded as the state constraint [136–139]. Simulations and experimental results reveal that MPC can prevent the sprung or unsprung masses from hitting the stopper. Compared to active suspension systems, MPC for semi-active control faces many difficulties. A main problem is the question of how to take the time-varying damper force constraint into consideration. Gorden et al. investigated the influence of horizon length on a semi-active suspension MPC controller [128]. Canale et al. then provided a fast MPC algorithm by using the piecewise affine (PWA) method, and a comparison with skyhook and clipped optimal control was then carried out. Simulation results showed that the proposed method can provide almost the same damping force compared to the traditional MPC algorithm. In addition, better ride comfort and handling capacity can be obtained [140]. Ahmed et al. performed semi-active MPC control for half and full car models, and a frequency-based controller was proposed [36, 141]. Giorgetti et al. compared clipped optimal, steepest gradient algorithm, optimal, and MPC controllers. The simulation results revealed that MPC performed better with increasing horizon length [142, 143].

The most important concepts of MPC include the predictive model, receding optimization, and feedback compensation. Figure 6.1 shows how MPC works, and the three concepts can be explained as follows.

1. Predictive model. The model for MPC can be any one of the following three categories [144, 145]: finite impulse response models, transfer function models, and state space-based models.

2. Receding optimization. MPC can be viewed as an open loop control problem with a finite horizon length. It can be seen from Figure 6.1 that at step t, the system calculates the states and the open loop control inputs until step $t + T_c$. The control input is the first open loop input at time, t. The controller repeats the above process at the next step, which is the primary difference with other algorithms.

3. Feedback compensation. The open loop control input cannot ensure the optimality because of the presence of disturbances and model uncertainties. MPC uses feedback compensation to improve its performance. In this chapter, state measurement is used for compensation.

The state space equation of a system can be generalized as follows:

$$\mathbf{x}(t + \Delta t) = \mathbf{A}\mathbf{x}(t) + \mathbf{B}\mathbf{u}(t) + \mathbf{\Gamma}\mathbf{w}(t)$$
$$\mathbf{y}(t) = \mathbf{C}\mathbf{x}(t) + \mathbf{D}\mathbf{u}(t). \tag{6.1}$$

The objective function of the above equation for $t \sim T_p$ can be expressed as:

$$J = \sum_{i=1}^{N_p} \mathbf{y}^T(t + i\Delta t)\mathbf{Q}\mathbf{y}(t + i\Delta t) + \sum_{j=0}^{N_c} \mathbf{u}^T(t + j\Delta t)\mathbf{R}\mathbf{u}(t + j\Delta t), \tag{6.2}$$

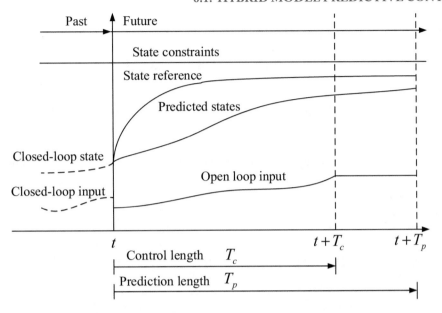

Figure 6.1: MPC algorithm.

where $\mathbf{Q}$ and $\mathbf{R}$ are the weighting matrices, and N_c and N_p are the control length and the prediction length, respectively.

Based on the state space equation, the estimated output can be written as:

$$\hat{\mathbf{y}} = \mathbf{\Lambda}\mathbf{x}(k) + \mathbf{\Gamma}_u\hat{\mathbf{u}} + \mathbf{\Gamma}_v\hat{\mathbf{w}}, \tag{6.3}$$

where $\hat{\mathbf{y}} = [\mathbf{y}(t + \Delta t), \ldots, \mathbf{y}(t + N_P)]^T$, $\hat{\mathbf{u}} = [\mathbf{u}(t), \mathbf{u}(t + \Delta t), \ldots, \mathbf{u}(t + N_C)]^T$, $\mathbf{\Lambda} \in \mathfrak{R}_{N_P \times 1}$, $\mathbf{\Gamma}_u \in \mathfrak{R}_{N_P \times (N_c+1)}$, $\mathbf{\Gamma}_v \in \mathfrak{R}_{N_P \times N_P}$, and $\hat{\mathbf{w}} = [\mathbf{w}(t), \mathbf{w}(t + \Delta t), \ldots, \mathbf{w}(t + N_P - 1)]^T$.
System output from step t to step $t + T_p$ can be calculated as follows:

$$\mathbf{y}(t + 1) = \mathbf{CA}\mathbf{x}(t) + \mathbf{CB}\mathbf{u}(t) + \mathbf{D}\mathbf{u}(t + \Delta t) + \mathbf{C}\mathbf{\Gamma}\mathbf{w}(t)$$
$$\mathbf{y}(t + 2\Delta t) = \mathbf{CA}^2\mathbf{x}(t) + \mathbf{CAB}\mathbf{u}(t) + \mathbf{CB}\mathbf{u}(t + \Delta t) + \mathbf{D}\mathbf{u}(t + 2\Delta t)$$
$$+ \mathbf{CA}\mathbf{\Gamma}\mathbf{w}(t) + \mathbf{C}\mathbf{\Gamma}\mathbf{w}(t + \Delta t)$$
$$\vdots$$
$$\mathbf{y}(t + N_P) = \mathbf{CA}\mathbf{x}(t) + \ldots + \mathbf{CA}^{N_P}\mathbf{x}(t + N_P) + \mathbf{CB}\mathbf{u}(t) + \ldots$$
$$+ \mathbf{D} + \left(\sum_{i=1}^{N_P-N_C} \mathbf{CA}^{i-1}\mathbf{B}\right)\mathbf{u}(t + N_C) + \ldots$$
$$+ \mathbf{C}\mathbf{\Gamma}\mathbf{w}(t) + \ldots + \mathbf{C}\mathbf{\Gamma}\mathbf{w}(t + N_P),$$

where $\mathbf{\Lambda}$, $\mathbf{\Gamma}_u$, and $\mathbf{\Gamma}_v$ are:

$$
\mathbf{\Lambda} = \begin{bmatrix} \mathbf{CA} \\ \mathbf{CA}^2 \\ \vdots \\ \mathbf{CA}^{N_P} \end{bmatrix}, \mathbf{\Gamma}_u = \begin{bmatrix} \mathbf{CB} & \mathbf{D} & 0 & \cdots & 0 \\ \mathbf{CAB} & \mathbf{CB} & \mathbf{D} & \cdots & 0 \\ \vdots & \vdots & \vdots & \vdots & \vdots \\ \mathbf{CA}^{N_C-1}\mathbf{B} & \mathbf{CA}^{N_C-2}\mathbf{B} & \cdots & \cdots & \mathbf{D} \\ \vdots & \vdots & \vdots & \vdots & \vdots \\ \mathbf{CA}^{N_P-1}\mathbf{B} & \mathbf{CA}^{N_P-2}\mathbf{B} & \cdots & \cdots & \mathbf{D} + \sum_{i=1}^{N_P-N_C} \mathbf{CA}^{i-1}\mathbf{B} \end{bmatrix},
$$

$$
\mathbf{\Gamma}_v = \begin{bmatrix} \mathbf{C\Gamma} & 0 & \cdots & \cdots & \cdots & 0 \\ \mathbf{CA\Gamma} & \mathbf{C\Gamma} & 0 & \cdots & \cdots & 0 \\ \vdots & \vdots & \vdots & \vdots & \vdots & \vdots \\ \mathbf{CA}^{N_C}\mathbf{\Gamma} & \mathbf{CA}^{N_C-1}\mathbf{\Gamma} & \cdots & \mathbf{C\Gamma} & \cdots & 0 \\ \vdots & \vdots & \vdots & \vdots & \vdots & \vdots \\ \mathbf{CA}^{N_P-1}\mathbf{\Gamma} & \mathbf{CA}^{N_P-2}\mathbf{\Gamma} & \cdots & \mathbf{CA}^{N_P-N_C-2}\mathbf{\Gamma} & \cdots & \mathbf{C\Gamma} \end{bmatrix}.
$$

Taking Eq. (6.3) into Eq. (6.2), the objective function can be rewritten as:

$$
\min_u J = \mathbf{y}^T \mathbf{Q}\mathbf{y} + \mathbf{u}^T \mathbf{R}\mathbf{u}
$$
$$
= (\mathbf{\Lambda}\mathbf{x}(t) + \mathbf{\Gamma}_u\mathbf{u} + \mathbf{\Gamma}_v\mathbf{w})^T \mathbf{Q}(\mathbf{\Lambda}\mathbf{x}(t) + \mathbf{\Gamma}_u\mathbf{u} + \mathbf{\Gamma}_v\mathbf{w}) + \mathbf{u}^T \mathbf{R}\mathbf{u}, \tag{6.4}
$$

which can be further expressed as:

$$
\min_u J = (\mathbf{\Lambda}\mathbf{x}(t) + \mathbf{\Gamma}_u\mathbf{u} + \mathbf{\Gamma}_v\mathbf{w})^T \mathbf{Q}(\mathbf{\Lambda}\mathbf{x}(t) + \mathbf{\Gamma}_u\mathbf{u} + \mathbf{\Gamma}_v\mathbf{w}) + \mathbf{u}^T \mathbf{R}\mathbf{u}
$$
$$
= \mathbf{x}^T \mathbf{\Lambda}^T \mathbf{Q}\mathbf{\Lambda}\mathbf{x} + \mathbf{u}^T \left(\mathbf{\Gamma}_u^T \mathbf{Q}\mathbf{\Gamma}_u + \mathbf{R}\right)\mathbf{u} + \mathbf{w}^T \mathbf{\Gamma}_v^T \mathbf{Q}\mathbf{\Gamma}_v\mathbf{w} \tag{6.5}
$$
$$
+ 2\mathbf{x}^T \mathbf{\Lambda}^T \mathbf{Q}\mathbf{\Gamma}_v\mathbf{w} + 2\mathbf{x}^T \mathbf{\Lambda}^T \mathbf{Q}\mathbf{\Gamma}_u\mathbf{u} + 2\mathbf{w}^T \mathbf{\Gamma}_v^T \mathbf{Q}\mathbf{\Gamma}_u\mathbf{u}.
$$

By removing the terms without the control force $\mathbf{u}$, Eq. (6.5) can be written in the following form to derive the minimum $\mathbf{u}$:

$$
\min_\mathbf{u} J = \mathbf{u}^T \left(\mathbf{\Gamma}_u^T \mathbf{Q}\mathbf{\Gamma}_u + \mathbf{R}\right)\mathbf{u} + 2\mathbf{x}^T \mathbf{\Lambda}^T \mathbf{Q}\mathbf{\Gamma}_u\mathbf{u} + 2\mathbf{w}^T \mathbf{\Gamma}_v^T \mathbf{Q}\mathbf{\Gamma}_u\mathbf{u}. \tag{6.6}
$$

The optimal trajectory can then be obtained by calculating $\partial \mathbf{J}/\partial \mathbf{u}$ when there is no constraint. The optimal control force is the same as in optimal feedback control:

$$
\mathbf{u}_{opt} = -\left(\mathbf{\Gamma}_u^T \mathbf{Q}\mathbf{\Gamma}_u + \mathbf{R}\right)^{-1} \mathbf{\Gamma}_u^T \mathbf{Q}(\mathbf{\Lambda}\mathbf{x}(t) + \mathbf{\Gamma}_v\mathbf{w}). \tag{6.7}
$$

Assuming that there is an output constraint in the system:

$$
\mathbf{L} \leq \mathbf{y} \leq \mathbf{U}. \tag{6.8}
$$

Equation (6.3) can be rewritten as:

$$\mathbf{L} - \boldsymbol{\Gamma}_u \hat{\mathbf{u}} - \boldsymbol{\Gamma}_v \hat{\mathbf{w}} < \boldsymbol{\Lambda}\mathbf{x}(k) < \mathbf{U} - \boldsymbol{\Gamma}_u \hat{\mathbf{u}} - \boldsymbol{\Gamma}_v \hat{\mathbf{w}} \tag{6.9}$$

$$\begin{bmatrix} \boldsymbol{\Gamma}_u \\ -\boldsymbol{\Gamma}_u \end{bmatrix} \mathbf{u} \leq \begin{bmatrix} \mathbf{U} \\ -\mathbf{L} \end{bmatrix} + \begin{bmatrix} -\boldsymbol{\Lambda} & -\boldsymbol{\Gamma}_v \\ \boldsymbol{\Lambda} & \boldsymbol{\Gamma}_v \end{bmatrix} \begin{bmatrix} x(t) \\ w \end{bmatrix}. \tag{6.10}$$

The system is now a typical Quadratic Program (QP) problem, which can be expressed as [146]:

$$\min_u J = \mathbf{u}^T \left(\boldsymbol{\Gamma}_u^T \mathbf{Q} \boldsymbol{\Gamma}_u + \mathbf{R} \right) \mathbf{u} + 2\mathbf{x}^T \boldsymbol{\Lambda}^T \mathbf{Q} \boldsymbol{\Gamma}_u \mathbf{u} + 2\mathbf{w}^T \boldsymbol{\Gamma}_v^T \mathbf{Q} \boldsymbol{\Gamma}_u \mathbf{u}$$

$$\text{s.t.} \begin{bmatrix} \boldsymbol{\Gamma}_u \\ -\boldsymbol{\Gamma}_u \end{bmatrix} \mathbf{u} \leq \begin{bmatrix} \mathbf{U} \\ -\mathbf{L} \end{bmatrix} + \begin{bmatrix} -\boldsymbol{\Lambda} & -\boldsymbol{\Gamma}_v \\ \boldsymbol{\Lambda} & \boldsymbol{\Gamma}_v \end{bmatrix} \begin{bmatrix} x(t) \\ w \end{bmatrix}. \tag{6.11}$$

The KKT condition for this problem is given below:

$$\left(\boldsymbol{\Gamma}_u^T \mathbf{Q} \boldsymbol{\Gamma}_u + \mathbf{R} \right) \mathbf{u} + \mathbf{x}^T \boldsymbol{\Lambda}^T \mathbf{Q} \boldsymbol{\Gamma}_u + \begin{bmatrix} \boldsymbol{\Gamma}_u \\ -\boldsymbol{\Gamma}_u \end{bmatrix}^T \lambda = 0$$

$$\begin{bmatrix} \boldsymbol{\Gamma}_u \\ -\boldsymbol{\Gamma}_u \end{bmatrix} \mathbf{u} \leq \begin{bmatrix} \mathbf{U} \\ -\mathbf{L} \end{bmatrix} + \begin{bmatrix} -\boldsymbol{\Lambda} & -\boldsymbol{\Gamma}_v \\ \boldsymbol{\Lambda} & \boldsymbol{\Gamma}_v \end{bmatrix} \begin{bmatrix} x(t) \\ w \end{bmatrix} \leq 0 \tag{6.12}$$

$$\lambda^T \left(\begin{bmatrix} \boldsymbol{\Gamma}_u \\ -\boldsymbol{\Gamma}_u \end{bmatrix} \mathbf{u} \leq \begin{bmatrix} \mathbf{U} \\ -\mathbf{L} \end{bmatrix} + \begin{bmatrix} -\boldsymbol{\Lambda} & -\boldsymbol{\Gamma}_v \\ \boldsymbol{\Lambda} & \boldsymbol{\Gamma}_v \end{bmatrix} \begin{bmatrix} x(t) \\ w \end{bmatrix} \right) = 0$$

$$\lambda \geq 0.$$

This QP problem can be solved using the active set method (ASM) or the interior point method (IPM).

6.1.2 HYBRID SYSTEM

The previous section provides a brief introduction to MPC. Although it has been applied in active suspension systems, the traditional MPC cannot be directly used in semi-active suspensions due to the time-varying and state-dependent damping force constraint. This section first introduces the unique constraints in semi-active suspension. Next, the semi-active suspension is transformed into a hybrid system, which can then be controlled by HMPC.

Chapter 2 introduced and created an analytical model for a controllable damper. In the following section, a simplification is first performed to simplify the system and speed up the simulation. Previous research has revealed that the maximum rattle space velocity is approximately 0.6 m/s for road level ISO-C at a vehicle velocity of 40 km/h. This section simplifies the controllable damper using the equivalent dissipation energy with boundary velocity equaling to ±0.6 m/s. The simplified damper is shown in Figure 6.2. The upper and lower bounds for positive and negative rattle space velocity are given by y_1, y_2, y_3, and y_4, respectively. It can be seen that the boundaries depend on rattle space velocity. Therefore, the control force is state-dependent, and the traditional MPC as well as the QP solver cannot resolve such a problem. This section will introduce the concept of a hybrid system, which can be resolved by mixed integer programming (MIP).

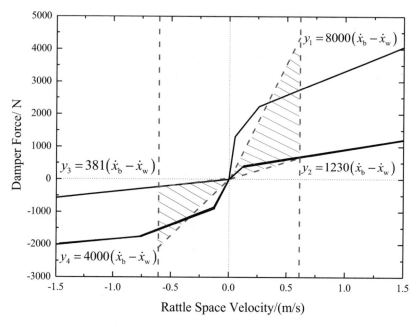

Figure 6.2: Simplified damper model.

In the previous chapters, the investigated systems are continuous physical processes, which are also called continuous variables dynamic system (CVDS). Compared to the CVDS, discrete event dynamic system (DEDS) combines both logical decision making and embedded control actions. This plays an important role in the multi-disciplinary design of many technological systems. Hybrid systems are developed based on DEDS, and the dynamics of both the continuous parts and discrete parts are combined to form the mathematical models. One distinct feature of a hybrid system is that all signals could be time-driven while others could be event-driven, even in an asynchronous manner. The conversion between the continuous parts and the discrete parts relies on the switch condition, which is typically a series of inequalities composed of continuous states. The hybrid system is graphically represented by Figure 6.3 [146].

Currently, there are many types of hybrid system models being studied. These include mixed logical dynamical (MLD), PWA, linear complementarity (LC), extended linear complementarity (ELC), and max-min-plus-scaling (MMPS) models. Different types of hybrid system

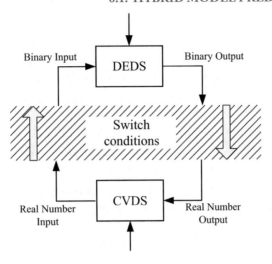

Figure 6.3: Hybrid system.

models can be mutually transformed under certain conditions. This is shown in Figure 6.4 [147]. Here, the star (*) indicate that the transformation requires conditions.

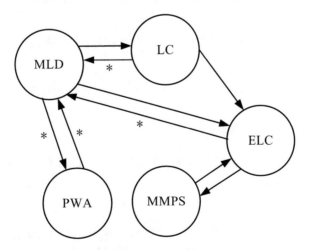

Figure 6.4: The link between different hybrid system models.

6.1.3 HYBRID MODEL PREDICTIVE CONTROL

PWA system partitions the extended state space into several polyhedral regions and describes each individual region by a distinct affine state equation. The discrete model of a PWA system can be depicted as:

$$x(t + \Delta t) = A^{i(t)}x(t) + B^{i(t)}u(t) + f^{i(t)}$$
$$y(t) = C^{i(t)}x(t) + D^{i(t)}u(t) + g^{i(t)} \tag{6.13}$$
$$H^{i(t)}x(t) + J^{i(t)}u(t) \leq K^{i(t)},$$

where $x \in \Re^{n_c} \times \{0,1\}^{n_l}$ is system state, $u \in \Re^{m_c} \times \{0,1\}^{m_l}$ is system input, $y \in \Re^{p_c} \times \{0,1\}^{p_l}$ is system output, and $i(t) \in \aleph = \{1, \ldots, s\}$ is used to indicate the current region. Each inequality in Eq. (6.13) defines a polyhedron $C^i = \left\{[x\,u]^T \in \Re^{m+n} : H^i x + J^i u \leq K^i\right\}$, where $m = m^c + m^l$ and $n = n^c + n^l$. The superscript l represents a binary value. Further, if the PWA system contains the following state and input constraint:

$$Ex(t) + Lu(t) \leq M, \tag{6.14}$$

which can be rewritten as:

$$x(t + \Delta t) = A^i x(t) + B^i u(t) + f^i \quad \text{if } [x(t)\,u(t)]^T \in \tilde{C}^i, \tag{6.15}$$

where $\{\tilde{C}^i\}_{i=1}^s$ is the new polyhedral region defined in space $\Re^{m+n}$ by considering C^i in Eq. (6.13) and the constraint in Eq. (6.14). Define system objective function as follows:

$$J(x(0), U_0) = \|Px_N\|_p + \sum_{k=0}^{N-1} \|Qx_k\|_p + \|Ru_k\|_p, \tag{6.16}$$

where $\|\cdot\|_p$ represents p-norm. Equation (6.13) now becomes a finite time optimal control problem:

$$J_0^*(x(0)) = \min_{U_0} J_0(x(0), U_0)$$
$$\text{s.t.} \begin{cases} x(t + \Delta t) = A^i x(t) + B^i u(t) + f^i \quad \text{if } [x(t)\,u(t)]^T \in \tilde{C}^i \\ x_N \in \chi_f \\ x_0 = x(0), \end{cases} \tag{6.17}$$

where $U_0 = \left[u_0', \ldots, u_{N-1}'\right]^T \in \Re^{m_c N} \times \{0,1\}^{m_l N}$ is the control input, N is control horizon, and χ_f is the terminal region. The state region χ_k at step k can then be given as:

$$\chi_k = \left\{ x \in \Re^{n_c} \times \{0,1\}^{n_l} \,\middle|\, \begin{matrix} \exists u \in \Re^{m_c} \times \{0,1\}^{m_l} \\ \exists x \in \Re^{n_c} \times \{0,1\}^{n_l} \\ [x\,u]^T \in \tilde{C}^i \\ A^i x(t) + B^i u(t) + f^i \in \chi_{k+1} \end{matrix} \right\}, \quad k = 0, 1, \ldots, N-1,$$

$$\chi_N = \chi_f. \tag{6.18}$$

Equation (6.18) implies that for any initial condition $x_i \in \chi_i$, there exists a control sequence $U_i = [u_i', \ldots, u_{N-1}']^T$, which can ensure $x_N \in \chi_f$ at step N.

Lemma 6.1   *A state feedback control law in the PWA form can be given for the problem stated in Eqs. (6.16) and (6.17) if $p = 2$ [146]:*

$$u_k^*(x(k)) = F_k^i x(k) + g_k^i \quad if x(k) \in \mathfrak{R}_k^i, \tag{6.19}$$

where $\mathfrak{R}_k^i$, $i = 1, \ldots, N_k$ is a partition of the set χ_k, and F_k^i and g_k^i are the weighting matrices in i th region. The polyhedral regions can be found offline, and only Eq. (6.19) is required to be resolved online.

The proof of the above lemma can be found in [146].

As stated in Lemma 6.1, the PWA system optimal control requires the computation of the control sequence in the form of the PWA offline. This process is time-consuming and the large computation time for the increased control horizon is a significant drawback. Bemporad et al. proposed the MLD model and the corresponding MID method to remedy this issue [148]. For the PWA model given by Eq. (6.15), its equivalent MLD model can be written as [147, 149]:

$$J_0^*(x(0)) = \min_{U_0} J_0(x(0), U_0)$$

$$\text{s.t.} \begin{cases} x(t+1) = Ax(t) + B_1 u(t) + B_2 \delta(t) + B_3 z(t) \\ y(t) = Cx(t) + D_1 u(t) + D_2 \delta(t) + D_3 z(t) + D_5 \\ E_2 \delta_k + E_3 z_k \leq E_1 u_k + E_4 x_k + E_5 \\ x_N \in \chi_f \\ x_0 = x(0), \end{cases} \tag{6.20}$$

where $x \in \mathfrak{R}^{n_c} \times \{0, 1\}^{n_l}$ is system state, $u \in \mathfrak{R}^{m_c} \times \{0, 1\}^{m_l}$ is system input, and $y \in \mathfrak{R}^{p_c} \times \{0, 1\}^{p_l}$ is system output. $\delta \in \{0, 1\}^{r_l}$ and $z \in \mathfrak{R}^{r_c}$ are auxiliary logical and continuous variables. The objective function can be given as:

$$J(x(t), U_0) = \|Px_N\|_p + \sum_{k=t}^{t+N} \|Qx_k\|_p + \|Ru_k\|_p + \|Q_\delta \delta_k\|_p + \|Q_z z_k\|_p. \tag{6.21}$$

The MIP problem corresponding to Eqs. (6.20) and (6.21) can be formulated in the following way:

$$\min_\varepsilon \varepsilon^T H_1 \varepsilon + \varepsilon^T H_2 x(t) + x(t)^T H_3 x(t) + c_1^T \varepsilon + c_2^T x(t) + c$$

$$\text{s.t. } G\varepsilon \leq w + Sx(t), \tag{6.22}$$

where $\varepsilon = [\varepsilon_c^T, \varepsilon_l^T]$ is a set combining continuous and logical variables. The first three terms in Eq. (6.21) are 2-norm, which represents mixed integer quadratic programming (MIQP). The last two terms can be either 1-norm or ∞-norm, which correspond to mixed integer linear programming (MILP). The MIP problem can be solved by one of the following three methods:

1. enumerate all logical variables in set ε_l, and solve as many MIP problems as there are elements in ε_l;

2. perform mixed integer optimization using the branch and bound method [150]; and

3. transform the problem into a multivariable programming problem by regarding the initial condition $x(0)$ as a variable [151].

6.2   OPTIMAL PREDICTIVE CONTROL

In Section 6.1, we introduced the concept of HMPC, and how to resolve the MIP problem. In order to compare the HMPC with other predictive algorithms, this section introduces optimal predictive control and how to use it to compensate for road disturbances in suspension system.

One common method to deal with the state-dependent force constraint in semi-active suspension systems is to apply the constraint in the fully active control force, which is also called clipped optimal control (COC) [152]. Based on this concept, this part introduces two lemmas for quarter and half vehicle models, whose proof can be found in [127].

6.2.1   PREDICTIVE CONTROL FOR QUARTER VEHICLE MODEL

For the model shown in Figure 5.9b, an objective function is given below:

$$J_{quarter} = \lim_{T \to \infty} \frac{1}{T} E \left\{ \int_0^T \left[q_1 \ddot{x}_b^2 + q_2 (x_b - x_w)^2 + q_3 (x_w - x_r)^2 + rU^2 \right] dt \right\}, \qquad (6.23)$$

where U is the unconstrained optimal force. The above equation can be rewritten as:

$$J_{quarter} = \lim_{T \to \infty} \frac{1}{T} E \left\{ \int_0^T \left[\mathbf{x}^T \mathbf{Q}_{1q} \mathbf{x} + 2\mathbf{x}^T \mathbf{N}_q U + U^T \mathbf{R}_q U + 2\mathbf{x}^T \mathbf{Q}_{12q} \mathbf{w} + \mathbf{w}^T \mathbf{Q}_{2q} \mathbf{w} \right] dt \right\},$$
$$(6.24)$$

where $\mathbf{Q}_{1q}$, $\mathbf{Q}_{12q}$, and $\mathbf{Q}_{2q}$ are time invariant symmetric positive semidefinite matrices. Other matrices are as follows:

$$\mathbf{Q}_{1q} = \begin{bmatrix} -\dfrac{k_s^2}{m_b^2} + 1 & 0 & 0 & 0 \\ 0 & 1 & 0 & 0 \\ 0 & 0 & 0 & 0 \\ 0 & 0 & 0 & 0 \end{bmatrix}, \ \mathbf{N}_q = \begin{bmatrix} -\dfrac{2k_s}{m_b^2} \\ 0 \\ 0 \\ 0 \end{bmatrix}, \ \mathbf{R}_q = \dfrac{1}{m_b^2} + r, \ \mathbf{Q}_{12q} = \mathbf{Q}_{2q} = 0.$$

Lemma 6.2 *Assuming the predictive length is t_p, for road profile $w(\sigma)$, $\sigma \in [t, t_p]$, the optimal control force $F_{d-quarter} = f[x(t), w(\sigma), \sigma \in [t, t_p]]$ can be derived as follows.*

Define the following variables:

$$\mathbf{A}_{nq} = \mathbf{A}_s - \mathbf{B}_s \mathbf{R}^{-1} \mathbf{N}^T, \quad \mathbf{Q}_{nq} = \mathbf{Q}_{1q} - \mathbf{N}_q \mathbf{R}_q^{-1} \mathbf{N}_q^T,$$

$\mathbf{Q}_{nq}$ *can be decomposed into* $\mathbf{Q}_{nq} = \mathbf{Z}_q^T \mathbf{Z}_q$ *if it is nonnegative definite, where* $\mathbf{Z}_q$ *is a square matrix. If* $(\mathbf{A}_{nq}, \mathbf{B}_{sq})$ *is stabilizable and* $(\mathbf{A}_{nq}, \mathbf{Z}_q)$ *is detectable,* $F_{d-quarter}$ *can be calculated as:*

$$F_{d-quarter} = -\mathbf{R}_q^{-1} \left[(\mathbf{N}_q^T + \mathbf{B}_s^T \mathbf{P}_q) \mathbf{x}(t) + \mathbf{B}_s^T \mathbf{r}_q(t) \right], \tag{6.25}$$

where $\mathbf{P}_q$ *is the solution to the following Riccati equation:*

$$\mathbf{P}_q \mathbf{A}_{nq} + \mathbf{A}_{nq}^T \mathbf{P}_q - \mathbf{P}_q \mathbf{B}_s \mathbf{R}_q^{-1} \mathbf{B}_s^T \mathbf{P}_q + \mathbf{Q}_{nq} = 0. \tag{6.26}$$

The feedforward term $\mathbf{r}_q(t)$ *in Eq. (6.25) can be written in the form:*

$$\mathbf{r}_q(t) = \int_0^{t_p + t_d} e^{\mathbf{A}_{cq}^T \sigma} \left(\mathbf{P}_q \mathbf{\Gamma}_s + \mathbf{Q}_{12} \right) w_1(t + \sigma) \, d\sigma, \tag{6.27}$$

where $\mathbf{A}_{cq} = \mathbf{A}_{nq} - \mathbf{B}_h \mathbf{R}_q^{-1} \mathbf{B}_h^T \mathbf{P}_q.$

6.2.2 PREDICTIVE CONTROL FOR A HALF VEHICLE MODEL

This section investigates the predictive optimal control for a half vehicle model. In this part, we use a wheelbase algorithm to improve system dynamical performance. The road profile is estimated at the front axle using the method proposed in Section 4.2, and a predictive controller is then formulated and synthesizes the damper force at the rear axle. A half vehicle model is graphically depicted in Figure 6.5, and the dynamics equations are given below:

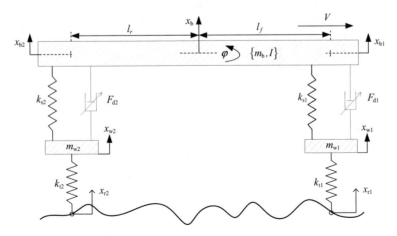

Figure 6.5: Half vehicle model.

$$\begin{cases} m_b \ddot{x}_b + k_{s1}(x_{b1} - x_{w1}) + F_{d1} + k_{s2}(x_{b2} - x_{w2}) + F_{d2} = 0 \\ I\ddot{\varphi} + l_f[k_{s1}(x_{b1} - x_{w1}) + F_{d1}] - l_r[k_{s2}(x_{b2} - x_{w2}) + F_{d2}] = 0 \\ m_{w1}\ddot{x}_{w1} + k_{s1}(x_{w1} - x_{b1}) - F_{d1} + k_{t1}(x_{w1} - x_{r1}) = 0 \\ m_{w2}\ddot{x}_{w2} + k_{s2}(x_{w2} - x_{b2}) - F_{d2} + k_{t2}(x_{w2} - x_{r2}) = 0, \end{cases} \tag{6.28}$$

where

$$\begin{cases} x_{b1} = x_b + l_f\varphi \\ x_{b2} = x_b - l_r\varphi. \end{cases} \tag{6.29}$$

The states are selected as $\mathbf{X} = [x_{b1} - x_{w1}, x_{b2} - x_{w2}, x_{w1} - x_{r1}, x_{w2} - x_{r2}, \dot{x}_{b1}, \dot{x}_{b2}, \dot{x}_{w1}, \dot{x}_{w2}]^T$, and the state space equation is described by

$$\dot{\mathbf{X}} = \mathbf{A}_h\mathbf{X} + \mathbf{B}_h\mathbf{U} + \boldsymbol{\Gamma}_h\boldsymbol{\omega}. \tag{6.30}$$

The matrices are given as:

$$\mathbf{A}_h = \begin{bmatrix} 0 & 0 & 0 & 0 & 1 & 0 & -1 & 0 \\ 0 & 0 & 0 & 0 & 0 & 1 & 0 & -1 \\ 0 & 0 & 0 & 0 & 0 & 0 & 1 & 0 \\ 0 & 0 & 0 & 0 & 0 & 0 & 0 & 1 \\ -k_{s1}a_1 & -k_{s2}a_2 & 0 & 0 & 0 & 0 & 0 & 0 \\ -k_{s1}a_2 & -k_{s2}a_3 & 0 & 0 & 0 & 0 & 0 & 0 \\ \dfrac{k_{s1}}{m_{w1}} & 0 & -\dfrac{k_{t1}}{m_{w1}} & 0 & 0 & 0 & 0 & 0 \\ 0 & \dfrac{k_{s2}}{m_{w2}} & 0 & -\dfrac{k_{t2}}{m_{w2}} & 0 & 0 & 0 & 0 \end{bmatrix}$$

$$\mathbf{B}_h = \begin{bmatrix} 0 & 0 \\ 0 & 0 \\ 0 & 0 \\ 0 & 0 \\ -a_1 & -a_2 \\ -a_2 & -a_3 \\ \dfrac{1}{m_{w1}} & 0 \\ 0 & \dfrac{1}{m_{w2}} \end{bmatrix} \qquad \boldsymbol{\Gamma}_h = \begin{bmatrix} 0 & 0 \\ 0 & 0 \\ -1 & 0 \\ 0 & -1 \\ 0 & 0 \\ 0 & 0 \\ 0 & 0 \\ 0 & 0 \end{bmatrix},$$

where $a_1 = \frac{1}{m_b} + \frac{l_f^2}{I}, a_2 = \frac{1}{m_b} - \frac{l_f l_r}{I}, a_1 = \frac{1}{m_b} + \frac{l_r^2}{I}.$

System output is then selected as $y = [\ddot{x}_b, \ddot{\varphi}, x_{b1} - x_{w1}, x_{b2} - x_{w2}, x_{w1} - x_{r1}, x_{w2} - x_{r2}]$, and the objective function is described by

$$
\begin{aligned}
J_{half} = \lim_{T \to \infty} \frac{1}{T} E \Bigg\{ \int_0^T \Bigg\{ &\begin{bmatrix} \ddot{x}_b \\ \ddot{\varphi} \end{bmatrix}^T \begin{bmatrix} \rho_1 & 0 \\ 0 & \rho_2 \end{bmatrix} \begin{bmatrix} \ddot{x}_b \\ \ddot{\varphi} \end{bmatrix} \\
&+ \begin{bmatrix} x_{b1} - x_{w1} \\ x_{b2} - x_{w2} \end{bmatrix}^T \begin{bmatrix} \rho_3 & 0 \\ 0 & \rho_4 \end{bmatrix} \begin{bmatrix} x_{b1} - x_{w1} \\ x_{b2} - x_{w2} \end{bmatrix} \\
&+ \begin{bmatrix} x_{w1} - x_{r1} \\ x_{w2} - x_{r2} \end{bmatrix}^T \begin{bmatrix} \rho_5 & 0 \\ 0 & \rho_6 \end{bmatrix} \begin{bmatrix} x_{w1} - x_{r1} \\ x_{w2} - x_{r2} \end{bmatrix} \\
&+ \begin{bmatrix} F_{d1} \\ F_{d2} \end{bmatrix}^T \begin{bmatrix} \rho_7 & 0 \\ 0 & \rho_8 \end{bmatrix} \begin{bmatrix} F_{d1} \\ F_{d2} \end{bmatrix} \Bigg\} dt \Bigg\},
\end{aligned}
\tag{6.31}
$$

which can be rewritten as

$$
\begin{aligned}
J_{half} = \lim_{T \to \infty} \frac{1}{T} E \\
\Bigg\{ \int_0^T \big[\mathbf{x}^T \mathbf{Q}_{1h} \mathbf{x} + 2\mathbf{x}^T \mathbf{N}_h \mathbf{U}_h + \mathbf{U}_h^T \mathbf{R}_h \mathbf{U}_h + 2\mathbf{x}^T \mathbf{Q}_{12h} \mathbf{w} + \mathbf{w}^T \mathbf{Q}_{2h} \mathbf{w} \big] dt \Bigg\},
\end{aligned}
\tag{6.32}
$$

where $\mathbf{Q}_{12h}$ and $\mathbf{Q}_{2h}$ are time invariant symmetric positive semidefinite matrices. Other matrices are as follows:

$$
\mathbf{Q}_{1h} = \begin{bmatrix} \mathbf{Q}_{11} & \mathbf{Q}_{12} \\ \mathbf{Q}_{21} & \mathbf{Q}_{22} \end{bmatrix}, \quad \text{where} \quad \mathbf{Q}_{12} = \mathbf{Q}_{21} = \mathbf{Q}_{22} = [0]_{4 \times 4},
$$

$$
\mathbf{N}_h = \begin{bmatrix} \mathbf{N}_{11} \\ \mathbf{N}_{21} \end{bmatrix}, \quad \text{where} \quad \mathbf{N}_{21} = [0]_{4 \times 2}, \mathbf{Q}_{12h} = [0]_{8 \times 2},
$$

$$
\mathbf{Q}_{2h} = [0]_{2 \times 2} \cdot \mathbf{Q}_{1h}.
$$

$$
\mathbf{Q}_{11} = \begin{bmatrix}
\dfrac{k_{s1}^2 \rho_1}{m_b^2} + \dfrac{l_f^2 k_{s1}^2 \rho_2}{I^2} + \rho_3 & \dfrac{k_{s1} k_{s2} \rho_1}{m_b^2} - \dfrac{l_f l_r k_{s1} k_{s2} \rho_2}{I^2} & 0 & 0 \\
\dfrac{k_{s1} k_{s2} \rho_1}{m_b^2} - \dfrac{l_f l_r k_{s1} k_{s2} \rho_2}{I^2} & \dfrac{k_{s2}^2 \rho_1}{m_b^2} + \dfrac{l_r^2 k_{s2}^2 \rho_2}{I^2} + \rho_4 & 0 & 0 \\
0 & 0 & \rho_5 & 0 \\
0 & 0 & 0 & \rho_6
\end{bmatrix}
$$

$$\mathbf{N}_{11} = \begin{bmatrix} \dfrac{k_{s1}\rho_1}{m_b^2} + \dfrac{l_f^2 k_{s1}\rho_2}{I^2} & \dfrac{k_{s1}\rho_1}{m_b^2} - \dfrac{l_f l_r k_{s1}\rho_2}{I^2} \\[2ex] \dfrac{k_{s2}\rho_1}{m_b^2} - \dfrac{l_f l_r k_{s2}\rho_2}{I^2} & \dfrac{k_{s2}\rho_1}{m_b^2} + \dfrac{l_r^2 k_{s2}\rho_2}{I^2} \\[2ex] 0 & 0 \\[1ex] 0 & 0 \end{bmatrix}$$

$$\mathbf{R}_h = \begin{bmatrix} \dfrac{\rho_1}{m_b^2} + \dfrac{l_f^2 \rho_2}{I^2} + \rho_7 & \dfrac{\rho_1}{m_b^2} - \dfrac{l_f l_r \rho_2}{I^2} \\[2ex] \dfrac{\rho_1}{m_b^2} - \dfrac{l_f l_r \rho_2}{I^2} & \dfrac{\rho_1}{m_b^2} + \dfrac{l_r^2 \rho_2}{I^2} + \rho_8 \end{bmatrix}.$$

Lemma 6.3 *Assuming the predictive length is t_p, for road profile $w(\sigma)$, $\sigma \in [t, t_p]$, the optimal control force $F_{d-half} = f[x(t), w(\sigma), \sigma \in [t, t_p]]$ can be derived as follows.*

Defining the following variables:

$$\mathbf{A}_{nh} = \mathbf{A}_h - \mathbf{B}_h \mathbf{R}_h^{-1} \mathbf{N}_h^T, \quad \mathbf{Q}_{nh} = \mathbf{Q}_{1h} - \mathbf{N}_h \mathbf{R}_h^{-1} \mathbf{N}_h^T,$$

$\mathbf{Q}_{nh}$ can be decomposed into $\mathbf{Q}_{nh} = \mathbf{Z}_h^T \mathbf{Z}_h$ if it is nonnegative definite, where $\mathbf{Z}_h$ is a square matrix. If $(\mathbf{A}_{nh}, \mathbf{B}_h)$ is stabilizable and $(\mathbf{A}_{nh}, \mathbf{Z}_h)$ is detectable, F_{d-half} can be calculated as:

$$F_{d-half} = -\mathbf{R}_h^{-1}\left[\left(\mathbf{N}_h^T + \mathbf{B}_h^T \mathbf{P}_h\right) \mathbf{x}(t) + \mathbf{B}_h^T \mathbf{r}_h(t) \right], \tag{6.33}$$

where $\mathbf{P}_h$ is the solution to the following Riccati equation:

$$\mathbf{P}_h \mathbf{A}_{nh} + \mathbf{A}_{nh}^T \mathbf{P}_h - \mathbf{P}_h \mathbf{B}_h \mathbf{R}_h^{-1} \mathbf{B}_h^T \mathbf{P}_h + \mathbf{Q}_{nh} = 0. \tag{6.34}$$

The feedforward term $\mathbf{r}_q(t)$ in Eq. (6.33) can be written in the form:

$$\mathbf{r}_h(t) = \int_0^{t_p + t_d} e^{\mathbf{A}_{ch}^T \sigma} \left(\mathbf{P}_h \mathbf{\Gamma}_h + \mathbf{Q}_{12h}\right) \begin{bmatrix} H(t_p - \sigma) w_1(t + \sigma) \\ w_1(t + \sigma - t_d) \end{bmatrix} d\sigma, \tag{6.35}$$

where t_d is the delay between the front and rear wheels, and $\mathbf{A}_{ch} = \mathbf{A}_{nh} - \mathbf{B}_h \mathbf{R}_h^{-1} \mathbf{B}_h^T \mathbf{P}_h$ is the closed loop system matrix. $H(t_p - \sigma)$ can be described by

$$H(t_p - \sigma) = \begin{cases} 1 & \text{if } \sigma \leq t_p \\ 0 & \text{if } \sigma > t_p. \end{cases}$$

6.3 SIMULATIONS

Here, the performance of the proposed controller is presented for quarter and half vehicle models.

6.3.1   SIMULATION SETTINGS

The settings that will be used in the simulations of the following parts are:

1. the vehicle is driving on road with a level of ISO-C at a speed 40 km/h;

2. the weighting matrices for different controllers are tuned to maintain the rattle space within 120 mm;

3. the prediction horizon is set as $N = 5, 20, 40$ for the HMPC;

4. for optimal predictive control, the predictive length is defined as $t_p = 0.005$ s, 0.01 s, 0.05 s, 0.15 s, 0.3 s, 0.5 s for the quarter vehicle, and $t_d = 0.0194$ s for the half vehicle model, which is calculated based on wheelbase and the velocity; and

5. in the following parts, COC and a system with a passive damper are used for comparison.

As discussed in Section 6.1, an MIP problem is required to be solved for HMPC. This book used the YALMIP toolbox to solve this problem using the Gurobi solver [153].

6.3.2   SIMULATION RESULTS FOR A QUARTER VEHICLE MODEL

For the quarter vehicle model, we assume that a camera/laser sensor is mounted on the sprung mass, which can measure the road profile at a length of l ahead of the car. The quarter vehicle model is shown in Figure 6.6 [154, 155].

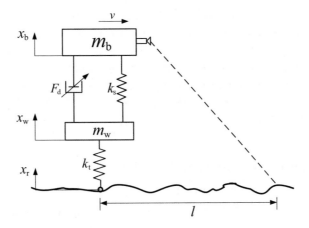

Figure 6.6: Predictive control for quarter vehicle model.

1. HMPC for the quarter vehicle model.

   For both the HMPC algorithms introduced in Section 6.1, the comparison of controller performance is tabulated in Table 6.1.

Table 6.1: Performance comparison of HMPC on quarter vehicle model

Responses	Passive	COC	HMPC (N=5)	HMPC (N=20)	HMPC (N=40)
RMS of SMA (m/s²)	0.866	0.791	0.751	0.736	0.729
RMS of TD (m)	0.00259	0.00248	0.00246	0.00245	0.00244

It can be seen from Table 6.1 that compared to the traditional semi-active controllers, i.e., COC, HMPC can improve system dynamic response by considering both state and force saturation. Additionally, the improvement becomes larger with the increasing prediction horizon. It should be noted, however, that such an improvement comes at the cost of additional computation complexity. Table 6.2 shows the required computation time for different prediction horizons.[1]

Table 6.2: Computation time of HMPC for different prediction horizon

Length	HMPC computation time (s)	MIP calculation time (s)
N=5	2.921	0.0089
N=20	4.059	0.0483
N=40	5.639	0.3312

The HMPC computation time in Table 6.2 is the total time for the controller generation, in which system states, output, and input are expressed as symbolic variables. The MIP calculation time represents the average time consumption for solving Eq. (6.22). It can be seen from Table 6.2 that longer prediction horizon length results in increased computation time. Compared to the HMPC computation time, the MIP calculation time is more sensitive to the prediction horizon. Therefore, it can be concluded that the improvement of HMPC controller performance can be achieved at the cost of additional computation expense.

2. Predictive control for the quarter vehicle model.

This section compares the predictive control with COC. Similar to the previous section, the performance comparison for different prediction lengths is given in Table 6.3.

It can be seen from Table 6.3 that the increased prediction horizon length can effectively improve the system dynamic response. However, the rate of improvement is not proportional to the increase in prediction length. From $t_p = 0.005$ s–$t_p = 0.15$ s, the improvement is more significant compared to the improvement from $t_p = 0.15$ s–$t_p = 0.5$ s. The computation time is also investigated and the results are shown in Table 6.4.

[1]All simulation results are obtained with i7 CPU@2.9 GHz and 8 GB RAM.

Table 6.3: Performance comparison of predictive control on quarter vehicle model

Responses	Passive	COC	PC (t_p = 0.005 s)	PC (t_p = 0.01 s)
RMS of SMA (m/s²)	0.866	0.791	0.791	0.731
RMS of TD (m)	0.00259	0.00248	0.00231	0.00218
	PC (t_p = 0.15 s)	PC (t_p = 0.3 s)	PC (t_p = 0.5 s)	
RMS of SMA (m/s²)	0.707	0.695	0.692	
RMS of TD (m)	0.00211	0.00208	0.00207	

Table 6.4: Computation time of predictive control for different prediction horizon

Prediction Length	Computation Time (s)
t_p = 0.005 s	0.00098
t_p = 0.01 s	0.0017
t_p = 0.15 s	0.0234
t_p = 0.3 s	0.0472
t_p = 0.5 s	0.0804

Table 6.4 reveals that, similar to HMPC, the performance improvement of predictive control is at the cost of computation burden.

In order to compare the performance of different controllers, a comparison of passive, different semi-active control strategies, and optimal predictive control for active suspension systems is presented here. The results are shown in Table 6.5 and Figures 6.7 and 6.8.

Table 6.5: Performance comparison for quarter vehicle model

Strategy	RMS of SMA (m/s²)	RMS of TD (m)
Passive	0.866	0.00259
COC	0.791 (8.66%↓)	0.00248 (4.25%↓)
HMPC, N=20	0.736 (15.1%↓)	0.00245 (5.41%↓)
PC, t_p = 0.15 s	0.707 (18.4%↓)	0.00211 (18.53%↓)
Active PC, t_p = 0.15 s	0.545 (37.1%↓)	0.00151 (41.71%↓)

Table 6.5 reveals that the application of all control strategies can improve vehicle vertical dynamics. Among all strategies, active PC performs the best with a 37% improvement in

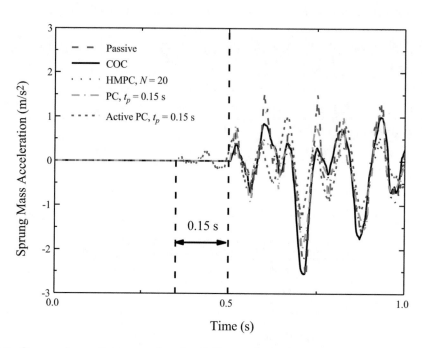

Figure 6.7: Comparison of ride comfort for different algorithms.

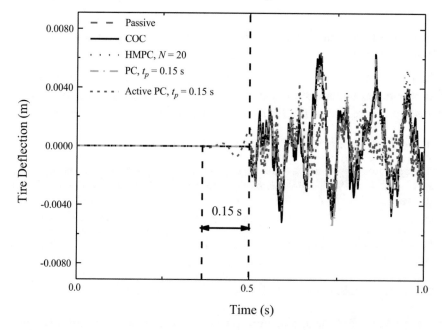

Figure 6.8: Comparison of handling capacity for different algorithms.

ride comfort and 42% improvement in handling capacity. For semi-active strategies, both HMPC and PC outperform the COC, which indicates that suspension performance can be enhanced by taking the damper force constraints into consideration. The time domain comparison is shown in Figures 6.7 and 6.8. For the first 0.5 s, the road excitation is assumed to be zero to illustrate the difference between active and semi-active predictive control algorithms.

From Figures 6.7 and 6.8, we can see that since the prediction horizon length for active PC is 0.15 s, the predictive active PC begins to generate force during 0.35–0.5 s. As for the semi-active predictive controllers, since dampers can only dissipate energy, no action is observed during this period. We can conclude that the differences of different control algorithms generally come from whether road information is used, and whether the damping force constraints are considered. We present the damper force and its corresponding bounds in the time domain in Figure 6.9.

It can be seen from Figure 6.9 that the controllable damper forces of all three semi-active control algorithms are inside the force bounds. For road level ISO-C, the required damper force is relatively small. Compared to other two algorithms, the HMPC damper force in Figure 6.9b is smaller and occurs at a lower frequency. Further, we define the power dissipation of the controllable damper as

$$P = F_d \left(\dot{x}_b - \dot{x}_w \right). \tag{6.36}$$

The dissipated powers of three algorithms are shown in Figure 6.10. The energy dissipation of HMPC is similar to that of COC. In contrast, the PC with a predictive horizon of $t_p = 0.15$ s dissipates less energy—approximately 65% of the energy is required compared to COC.

6.3.3 SIMULATION RESULTS FOR HALF VEHICLE MODEL

This section introduces the simulation results for the half vehicle model, and the wheelbase preview concept is used to obtain the road profile. This concept was first proposed by Louam et al. [156], in which it was assumed that the excitation of the rear wheel was the same as that of the front wheel with a delay determined by the vehicle velocity and wheelbase. In the part of this chapter, we use the algorithm proposed in Chapter 4 to estimate road profile at the front wheels and perform wheelbase preview control to improve vehicle's dynamic response. The parameters for the half vehicle model shown in Figure 6.5 are given in Table 6.6.

Note that for the passive suspension, the current for the damper is set to be 0.3 A for comparison. Next, we use a histogram to compare sprung mass acceleration, pitch acceleration, rattle space, and tire deflection for different controllers. The results are shown in Figure 6.11. The controllers shown in Figure 6.11 are described as follows.

- COC: two dampers with controlled by COC controller without preview.

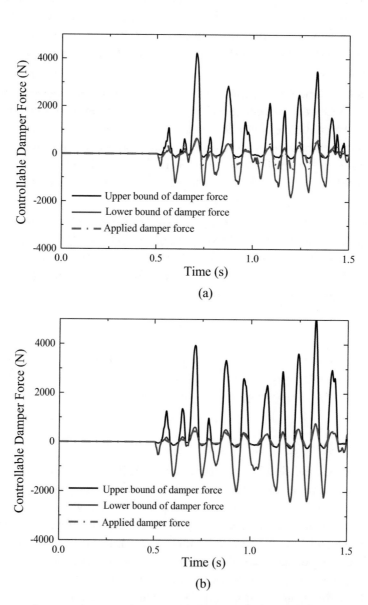

Figure 6.9: Damper force in the time domain: (a) COC and (b) HMPC, $N = 20$. (*Continues.*)

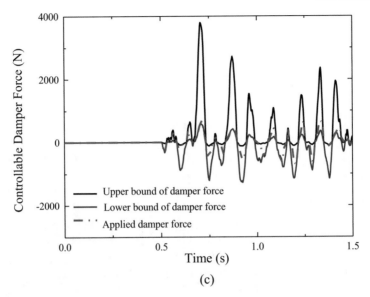

Figure 6.9: (*Continued.*) Damper force in the time domain: (c) PC, $t_p = 0.15$ s.

Table 6.6: Performance comparison for quarter vehicle model

Parameter	Value	Description
m_b(kg)	700	Sprung mass
I (kgm^2)	1,400	Pitch inertia
m_{w1}, m_{w2} (kg)	50	Unsprung mass
k_{s1}, k_{s2} (N/m)	20,000	Stiffness
k_{t1}, k_{t2} (N/m)	180,000	Tire stiffness
l_1, l_2 (m)	1.4	Front/Rear axle to C.G.
Current (A)	0.3	Passive damper control current

- SA-Wheelbase: passive damper in the front suspension, and predictive control for the rear.

- SA-HMPC: passive damper in the front suspension, and HMPC for the rear.

- A-Wheelbase: passive damper in the front suspension, and active optimal control for the rear.

- Active Preview: full active optimal control for both suspensions, and a predictive horizon of $t_p = 0.005$ s.

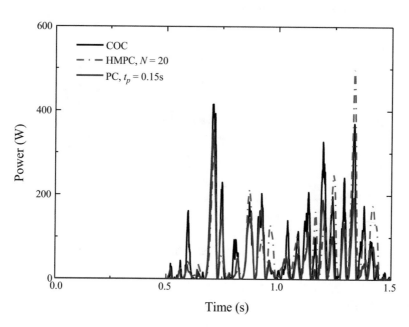

Figure 6.10: Energy dissipation for different semi-active controllers.

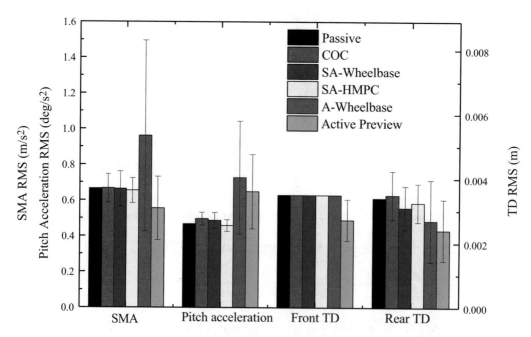

Figure 6.11: Wheelbase preview control for half vehicle model.

In Figure 6.11, the height of each bar represents the mean value of each response and the error bars show the performance variations caused by tuning controller parameters. Since the tire deflection is much smaller than the other two responses, the values of both front and rear tire deflections are shown on the right y-axis. According to Figure 6.11, the following conclusions can be drawn.

1. Improved sprung mass acceleration results in a smaller pitch acceleration response and vice versa. This indicates that these two responses are correlated.

2. Improved sprung mass acceleration causes an increase of rear tire deflection. This illustrates that the contradiction between ride comfort and handling capacity exists in the half vehicle model.

3. For all wheelbase preview algorithms, the variation of rear tire deflection has no effect on that of the front. The reason for this is the distance from the front axle to C.G. is the same with that of the rear axle.

For the full active case, all four responses can be improved by carefully selecting the control weighting matrix. Compared to the rear suspension, the regulating range of the front tire deflection is smaller. The reason for this is the prediction time for the rear suspension (0.0244 s) is much larger than that of the front suspension (0.005 s). This larger predictive horizon length enables more significant improvement for the rear suspension. Further, it can be seen that the tire deflection of active wheelbase preview control is almost the same as the fully active case, however, this is at the cost of increased sprung mass acceleration. For all of the semi-active control strategies, the wheelbase predictive control outperforms the other two controllers. The adjustable ranges of HMPC and COC are relatively small, but better performance can be obtained with HMPC.

According to the simulation results given in this section, the following conclusions can be drawn.

1. Compared to the traditional semi-active suspension control algorithms, HMPC can improve system dynamic performance since it can take system constraints into consideration by solving the MIQP.

2. If the road profile information is known in advance, predictive control can provide better performance. In the quarter vehicle model, the performance of HMPC with $N = 40$ is similar to that of predictive control with $t_p = 0.005$ s.

3. With the wheelbase preview control, the performance of the rear suspension of the half vehicle model is almost the same as the quarter vehicle model.

6.4   SUMMARY

This chapter introduced predictive semi-active suspension controllers to reduce the vibration caused by road disturbances. The concepts of model predictive and optimal predictive control were first introduced. Next, we presented HMPC to deal with the state-dependent damping force constraints. The system was written in the form of MLD, and the new system with the constraint was transformed into a MIP. This chapter used the YALMIP toolbox and the Gurobi solver to solve the above problem, and then applied traditional COC, HMPC, and PC on quarter and half vehicle models. In the quarter vehicle model, we assumed the road ahead could be measured by remote sensors. In the half vehicle model, the wheelbase preview concept was used, and we applied the road profile estimation method proposed in Chapter 4 to observe the road profile at the front suspension. Simulations for the two models were presented, and the results indicated that the HMPC could outperform the traditional optimal control algorithm since it accounted for the damper force constraints. Furthermore, it was shown that predictive control could provide better performance, and the performance of HMPC with $N = 40$ was similar to that of predictive control with $t_p = 0.005$ s in the quarter vehicle model.

CHAPTER 7

Conclusions

The main objectives of this book were to design algorithms for either statistical feature or time domain road estimation, and enhance vehicle dynamics performance based on the estimated results. Chapter 1 first introduced the current developments in road estimation, especially for system response based estimation algorithms. Road profile definitions and generation were comprehensively introduced in Chapter 2, in which the suspension model and controllable damper were also presented. Chapter 3 introduced a data-driven road classification algorithm, and the concept of candidate feature reduction to a "superior" feature set was carried out based on time-frequency domain analysis. Numerical simulation results revealed that the proposed algorithm was robust to variations in system parameters. The combination of the analysis and mutual information-based feature reduction methods can provide satisfactory classification accuracy. In Chapter 4, two model based road estimation algorithms were proposed for classification and profile estimation in the time domain. The classification algorithm was based on the transfer function from road excitation to UMA, and field test results validated the algorithm. For the time domain profile estimation algorithm, only measurable responses were used to formulate the input observer. Simulation results indicated that the proposed algorithm had higher accuracy compared to the traditional method. Chapter 5 demonstrated how the road level information could be used for semi-active suspension control. The analytical expressions of suspension responses w.r.t. road excitation were first introduced, and the suspension MOOP was then resolved using the NSGA-II approach. Simulation results showed effective vibration reduction for suspension systems. In Chapter 6, a HMPC was used to deal with the state-dependent damping force constraints that exist in semi-active suspension systems. This chapter used the YALMIP toolbox and the Gurobi solver, and then applied traditional COC, HMPC, and PC on both quarter and half vehicle models. The simulation results indicated that the HMPC could outperform the traditional optimal control algorithm since it accounted for damper force constraints.

References

[1] C. Hu, Y. Qin, H. Cao, et al. Lane keeping of autonomous vehicles based on differential steering with adaptive multivariable super-twisting control. *Mechanical Systems and Signal Processing*, 2018. DOI: 10.1016/j.ymssp.2018.09.011. 1

[2] Y. Qin, Z. Wang, C. Xiang, et al. A novel global sensitivity analysis on the observation accuracy of the coupled vehicle model. *Vehicle System Dynamics*, 1–22, 2018. DOI: 10.1080/00423114.2018.1517219.

[3] C. Hu, R. Wang, F. Yan, et al. Robust composite nonlinear feedback path-following control for independently actuated autonomous vehicles with differential steering. *IEEE Transactions on Transportation Electrification*, 2(3):312–321, 2016. DOI: 10.1109/tte.2016.2538183.

[4] X. Tang, D. Zhang, T. Liu, et al. Research on the energy control of a dual-motor hybrid vehicle during engine start-stop process. *Energy*, 2018. DOI: 10.1016/j.energy.2018.10.130.

[5] X. Tang, X. Hu, W. Yang, et al. Novel torsional vibration modeling and assessment of a power-split hybrid electric vehicle equipped with a dual-mass flywheel. *IEEE Transactions on Vehicular Technology*, 67(3):1990–2000, 2018. DOI: 10.1109/tvt.2017.2769084.

[6] T. Liu and X. Hu. A bi-level control for energy efficiency improvement of a hybrid tracked vehicle. *IEEE Transactions on Industrial Informatics*, 14(4):1616–1625, 2018. DOI: 10.1109/tii.2018.2797322.

[7] T. Liu, X. Hu, S. Li, et al. Reinforcement learning optimized look-ahead energy management of a parallel hybrid electric vehicle. *IEEE/ASME Transactions on Mechatronics*, 22(4):1497–1507, 2017. DOI: 10.1109/tmech.2017.2707338.

[8] T. Liu, Y. Zou, D. Liu, et al. Reinforcement learning of adaptive energy management with transition probability for a hybrid electric tracked vehicle. *IEEE Transactions on Industrial Electronics*, 62(12):7837–7846, 2015. DOI: 10.1109/tie.2015.2475419.

[9] X. Tang, J. Zhang, L. Zou, et al. Study on the torsional vibration of a hybrid electric vehicle powertrain with compound planetary power-split electronic continuous variable transmission. *Proc. of the Institution of Mechanical Engineers, Part C: Journal of Mechanical Engineering Science*, 228(17):3107–3115, 2014. DOI: 10.1177/0954406214526162.

[10] Y. Huang, H. Wang, A. Khajepour, et al. Model predictive control power management strategies for HEVs: A review. *Journal of Power Sources*, 341, 91–106, 2017. DOI: 10.1016/j.jpowsour.2016.11.106. 1

[11] D. Cao, X. Song, and M. Ahmadian. Editors' perspectives: Road vehicle suspension design, dynamics, and control. *Vehicle System Dynamics*, 49(1–2):3–28, 2011. DOI: 10.1080/00423114.2010.532223. 1

[12] W. Paterson. *Road Deterioration and Maintenance Effects: Models for Planning and Management*, Johns Hopkins University Press, Baltimore, MD, 1987. 1

[13] TRIP. *America's Roughest Rides and Strategies to Make our Roads Smoother*, Washington, DC, 2013. 1

[14] DGIP. EU road surfaces: Economic and safety impact of the lack of regular road maintenance, study. *Policy Department Structural and Cohesion Policies*, European Parliament, 2014. 1

[15] S. Gerwens. European Manifesto: Need for road maintenance. *Pavement Preservation and Recycling*, Summit, 2015. 1

[16] Transportation US. Long-term pavement performance program. 1

[17] S. Yang, L. Chen, and S. Li. *Dynamics of Vehicle-Road Coupled System*, Springer, 2015. DOI: 10.1007/978-3-662-45957-7. 1

[18] Z. Li, I. Kolmanovsky, E. Atkins, et al. Road disturbance estimation and cloud-aided comfort-based route planning. *IEEE Transactions on Cybernetics*, 47(11):3879–3891, 2017. DOI: 10.1109/tcyb.2016.2587673. 1, 7

[19] Z. Zhang, C. Sun, R. Bridgelall, et al. Application of a machine learning method to evaluate road roughness from connected vehicles. *Journal of Transportation Engineering, Part B: Pavements*, 144(4):04018043, 2018. DOI: 10.1061/jpeodx.0000074. 1

[20] R. Bridgelall, Y. Huang, Z. Zhang, et al. Precision enhancement of pavement roughness localization with connected vehicles. *Measurement Science and Technology*, 27(2):025012, 2016. DOI: 10.1088/0957-0233/27/2/025012. 1

[21] Z. Zhang, C. Sun, R. Bridgelall, et al. Road profile reconstruction using connected vehicle responses and wavelet analysis. *Journal of Terramechanics*, 80:21–30, 2018. DOI: 10.1016/j.jterra.2018.10.004. 1

[22] Z. Zhang, F. Deng, Y. Huang, et al. Road roughness evaluation using in-pavement strain sensors. *Smart Materials and Structures*, 24(11):115029, 2015. DOI: 10.1088/0964-1726/24/11/115029. 1

[23] D. Hrovat. Survey of advanced suspension developments and related optimal control applications. *Automatica*, 33(10):1781–1817, 1997. DOI: 10.1016/s0005-1098(97)00101-5. 1, 14

[24] K. Yi and B. Song. A new adaptive sky-hook control of vehicle semi-active suspensions. *Proc. of the Institution of Mechanical Engineers, Part D: Journal of Automobile Engineering*, 213(3):293–303, 1999. DOI: 10.1243/0954407991526874. 2, 5

[25] J. Zhao, P. Wong, X. Ma, et al. Design and analysis of an integrated SMC-TPWP strategy for a semi-active air suspension with stepper motor-driven GFASA. *Proc. Institution Mechanical Engineers Part I: Journal of Systems and Control Engineering*, 2018. DOI: 10.1177/0959651818778217.

[26] J. Zhao, P. Wong, Z. Xie, et al. Design and control of an automotive variable hydraulic damper using cuckoo search optimized PID method. *International Journal of Automation Technology*, 20, 2018. DOI: 10.1007/s12239-019-0005-z. 2

[27] Y. Qin, M. Dong, F. Zhao, et al. Comprehensive analysis for influence of controllable damper time delay on semi-active suspension control strategies. *Journal of Vibration and Acoustics Transactions of ASME*, 139(6):031006, 2017. DOI: 10.1115/1.4035700. 2, 13

[28] K. Hong, H. Sohn, and J. Hedrick. Modified skyhook control of semi-active suspensions: A new model, gain scheduling, and hardware-in-the-loop tuning. *Journal of Dynamic Systems, Measurement, and Control*, 124(1):158–167, 2002. DOI: 10.1115/1.1434265. 2, 6, 86

[29] G. Xue, H. Zhu, Z. Hu, et al. Pothole in the dark: Perceiving pothole profiles with participatory urban vehicles. *IEEE Transactions on Mobile Computing*, 16(5):1408–1419, 2017. DOI: 10.1109/tmc.2016.2597839. 2, 9

[30] Y. Qin, R. Langari, and L. Gu. The use of vehicle dynamic response to estimate road profile input in time domain. *ASME Dynamic System Control Conference (DSCC)*, San Antonio, TX, 2014. DOI: 10.1115/dscc2014-5978. 2, 7

[31] K. Mcghee. Automated pavement distress collection techniques. *Transportation Research Board*, 2004. 3

[32] H. Imine, Y. Delanne, and N. K. M'sirdi. Road profile input estimation in vehicle dynamics simulation. *Vehicle System Dynamics*, 44(4):285–303, 2006. DOI: 10.1080/00423110500333840. 3, 4, 5

[33] Acuity AR600/RP: Laser displacement sensor for road profilometry. https://www.Acuitylaser.Com/Docs/Ar600-Road-Profiling.Pdf 4

[34] Z. Yuan, X. Zhang, S. Liu, et al. Laser line recognition for autonomous road roughness measurement. *Proc. of the Cyber Technology in Automation, Control, and Intelligent Systems (Cyber)*, 2015. DOI: 10.1109/cyber.2015.7287977. 4

[35] M. Aki, T. Rojanaarpa, K. Nakano, et al. Road surface recognition using laser radar for automatic platooning. *IEEE Transactions on Intelligent Transportation Systems*, 17(10):2800–2810, 2016. DOI: 10.1109/tits.2016.2528892. 4

[36] M. Ahmed and F. Svaricek. Preview optimal control of vehicle semi-active suspension based on partitioning of chassis acceleration and tire load spectra. *Proc. of the Control Conference (ECC)*, 2014. DOI: 10.1109/ecc.2014.6862615. 5, 94

[37] A. Turnip and K. Hong. Road-frequency based optimization of damping coefficients for semi-active suspension systems. *International Journal of Vehicle Design*, 63(1):84–101, 2013. DOI: 10.1504/ijvd.2013.055493. 5

[38] D. Hrovat and W. Tseng. Adaptive active vehicle suspension system. US Patents. 1995. 5

[39] I. Fialho and G. Balas. Road adaptive active suspension design using linear parameter-varying gain-scheduling. *IEEE Transactions on Control Systems Technology*, 10(1):43–54, 2002. DOI: 10.1109/87.974337. 5

[40] D. Moustapha, V. Alessandro, A. Charara, et al. Estimation of road profile for vehicle dynamics motion: Experimental validation. *American Control Conference*, San Francisco, CA, 2011. DOI: 10.1109/acc.2011.5991595. 3, 5, 65

[41] J. Tudon-Martinez, S. Fergani, O. Sename, et al. Adaptive road profile estimation in semiactive car suspensions. *IEEE Transactions on Control Systems Technology*, 23(6):2293–2305, 2015. DOI: 10.1109/tcst.2015.2413937. 5

[42] A. Gonzalez, E. O'brien and K. Cashell. The use of vehicle acceleration measurements to estimate road roughness. *Vehicle System Dynamics*, 46(6):483–499, 2008. DOI: 10.1080/00423110701485050. 6

[43] S. Wang, S. Kodagoda, L. Shi, et al. Road-terrain classification for land vehicles: Employing an acceleration-based approach. *IEEE Vehicular Technology Magazine*, 12(3):34–41, 2017. DOI: 10.1109/mvt.2017.2656949. 6

[44] C. Gorges, R. Özt, and R. Liebich. Road classification for two-wheeled vehicles. *Vehicle System Dynamics*, 4:1–26, 2017. DOI: 10.1080/00423114.2017.1413197. 6

[45] Y. Qin, Z. Wang, C. Xiang, et al. Speed independent road classification strategy based on vehicle response: Theory and experimental validation. *Mechanical Systems and Signal Processing*, 117:653–666, 2019. DOI: 10.1016/j.ymssp.2018.07.035. 6, 45, 47, 57

[46] W. Fauriat, C. Mattrand, N. Gayton, et al. Estimation of road profile variability from measured vehicle responses. *Vehicle System Dynamics*, 53(5):585–605, 2016. DOI: 10.1080/00423114.2016.1145243. 6

[47] H. Ngwangwa, P. Heyns, H. Breytenbach, et al. Reconstruction of road defects and road roughness classification using artificial neural networks simulation and vehicle dynamic responses: Application to experimental data. *Journal of Terramechanics*, 53:1–18, 2014. DOI: 10.1016/j.jterra.2009.08.007. 6, 43

[48] C. Gorges, K. Öztürk, R. Liebich. Impact detection using a machine learning approach and experimental road roughness classification. *Mechanical Systems and Signal Processing*, 117:738–856, 2019. DOI: 10.1016/j.ymssp.2018.07.043. 6

[49] Y. Qin, R. Langari, Z. Wang, et al. Road excitation classification for semi-active suspension system with deep neural networks. *Journal of Intelligent and Fuzzy Systems*, 33(3):1907–1918, 2017. DOI: 10.3233/jifs-161860. 6

[50] P. Nitsche, M. Kammer. Comparison of machine learning methods for evaluating pavement roughness based on vehicle response. *Journal of Computing in Civil Engineering*, 28(4):04014015, 2012. DOI: 10.1061/(asce)cp.1943-5487.0000285. 6

[51] Y. Qin, C. Wei, X. Tang, et al. A novel nonlinear road profile classification approach for controllable suspension system: Simulation and experimental validation. *Mechanical Systems and Signal Processing*, 2018. DOI: 10.1016/j.ymssp.2018.07.015. 6

[52] L. Nguyen, K. Hong, and S. Park. Road-frequency adaptive control for semi-active suspension systems. *International Journal of Control, Automation, and Systems*, 8(5):1029–1038, 2010. DOI: 10.1007/s12555-010-0512-1. 6

[53] Z. Wang, M. Dong, L. Gu, et al. Influence of road excitation and steering wheel input on vehicle system dynamic responses. *Applied Sciences*, 7(6):570, 2017. DOI: 10.3390/app7060570. 6

[54] Z. Wang, M. Dong, Y. Qin, et al. Suspension system state estimation using adaptive Kalman filtering based on road classification. *Vehicle System Dynamics*, 55(3):371–398, 2017. DOI: 10.1080/00423114.2016.1267374. 15, 62

[55] Z. Wang, M. Dong, Y. Qin, et al. State estimation based on interacting multiple mode kalman filter for vehicle suspension system. *SAE Technical Paper*, 2017. DOI: 10.4271/2017-01-1480. 6

[56] M. Mahmoodabadi, A. Safaie, A. Bagheri, et al. A novel combination of particle swarm optimization and genetic algorithm for pareto optimal design of a five-degree of freedom vehicle vibration model. *Applied Soft Computing*, 13(5):2577–2591, 2013. DOI: 10.1016/j.asoc.2012.11.028. 6

[57] S. Kang, J. Kim, and G. Kim. Road roughness estimation based on discrete Kalman filter with unknown input. *Vehicle System Dynamics*, 2018. DOI: 10.1080/00423114.2018.1524151. 6

[58] Z. Wang, M. Dong, Y. Qin, et al. Road profile estimation for suspension system based on the minimum model error criterion combining with kalman filter. *Journal of Vibro-engineering*, 2017. 6

[59] D. Hassen, M. Miladi, M. Abbes, et al. Road profile estimation using the dynamic responses of the full vehicle model. *Applied Acoustics*, 2017. DOI: 10.1016/j.apacoust.2017.12.007. 7

[60] S. Kommuri, J. Rath, and K. Veluvolu. Sliding mode based observer-controller structure for fault-resilient control in DC servomotors. *IEEE Transactions on Industrial Electronics*, 99, 2017. DOI: 10.1109/tie.2017.2721883. 7

[61] J. Rath, K. Veluvolu, M. Defoort, et al. Higher-order sliding mode observer for estimation of tyre friction in ground vehicles. *IET Control Theory and Applications*, 8(6):399–408, 2014. DOI: 10.1049/iet-cta.2013.0593. 7

[62] J. Rath, K. Veluvolu, and M. Defoort. Simultaneous estimation of road profile and tire road friction for automotive vehicle. *IEEE Transactions on Vehicular Technology*, 64(10):4461–4471, 2015. DOI: 10.1109/tvt.2014.2373434. 7

[63] Z. Li, U. Kalabic, I. Kolmanovsky, et al. Simultaneous road profile estimation and anomaly detection with an input observer and a jump diffusion process estimator. *American Control Conference*, Boston, MA, 2016. DOI: 10.1109/acc.2016.7525160. 7

[64] C. Gohrle, A. Schindler, A. Wagner, et al. Road profile estimation and preview control for low-bandwidth active suspension systems. *IEEE/ASME Transactions on Mechatronics*, 20(5):2299–2310, 2015. DOI: 10.1109/tmech.2014.2375336. 7

[65] C. Huang, J. Lin, and C. Chen. Road-adaptive algorithm design of half-car active suspension system. *Expert Systems with Applications*, 37(6):4392–4402, 2010. DOI: 10.1016/j.eswa.2009.11.089. 7

[66] Z. Li, D. Filev, I. Kolmanovsky, et al. A new clustering algorithm for processing gps-based road anomaly reports with a mahalanobis distance. *IEEE Transactions on Intelligent Transportation Systems*, 18(7):1980–1988, 2017. DOI: 10.1109/tits.2016.2614350. 7

[67] ISO. Mechanical vibration-road surface profiles-reporting of measured data. *International Organization for Standardization*, 2016. DOI: 10.3403/30341600. 9

[68] J. Robson and C. Dodds. The description of road surface roughness. *Journal of Sound and Vibration*, 31:175–183, 1973. DOI: 10.1016/s0022-460x(73)80373-6. 9, 11, 14

[69] K. Bogsj, K. Podg, and I. Rychlik. Models for road surface roughness. *Vehicle System Dynamics*, 50(5):725–747, 2012. DOI: 10.1080/00423114.2011.637566. 9

[70] M. Sergio, P. S. Charles, S. Cristiano, et al. *Semi-Active Suspension Control Design for Vehicles*, Elsevier, 2010. DOI: 10.1016/c2009-0-63839-3.

[71] J. J. Rath, M. Defoort, H. R. Karimi, et al. Output feedback active suspension control with higher order terminal sliding mode. *IEEE Transactions on Industrial Electronics*, 2017. DOI: 10.1109/TIE.2016.2611587. 9

[72] D. Charles. Derivation of environment descriptions and test severities from measured road transportation data. *Journal of the IES*, 36(1):37–42, 1993. 9

[73] B. Bruscella, V. Rouillard, and M. Sek. Analysis of road surface profiles. *Journal of Transportation Engineering*, 125(1):55–59, 1999. DOI: 10.1061/(asce)0733-947x(1999)125:1(55). 9

[74] Z. Li, I. Kolmanovsky, E. Atkins, et al. Road anomaly estimation: Model based pothole detection. *American Control Conference*, 2015. DOI: 10.1109/acc.2015.7170915. 9

[75] M. Riveiro, M. Lebram, and M. Elmer. Anomaly detection for road traffic: A visual analytics framework. *IEEE Transactions on Intelligent Transportation Systems*, 2017. DOI: 10.1109/tits.2017.2675710.

[76] O. Kropac and P.Mucka. Specification of obstacles in the longitudinal road profile by median filtering. *Journal of Transportation Engineering*, 137(3):213–226, 2011. DOI: 10.1061/(asce)te.1943-5436.0000209.

[77] R. Wang, Y. Chuang, and C. Yi. A crowdsourcing-based road anomaly classification system. *Proc. of the Network Operations and Management Symposium*, 2016. DOI: 10.1109/apnoms.2016.7737244. 9

[78] P. Michelberger, L. Palkovics, and J. Bokor. Robust design of active suspension system. *International Journal of Vehicle Design*, 14(2/3):145–165, 1993. 11, 13, 14

[79] Z. Yi. Time domain model of road undulation excitation to vehicles. *Transactions of the Chinese Society of Agricultural Machinery*, 2004. DOI: 10.4028/www.scientific.net/AMR.216.96. 11

[80] Z. Guosheng, Z. Fang, S. Zhang, et al. White noise simulation for road roughness based on power function. *Automotive Engineering*, 30(1):44–47, 2008. 11, 13

[81] O. Kropac and P. Mucka. Effects of longitudinal road waviness on vehicle vibration response. *Vehicle System Dynamics*, 47(2):135–153, 2009. DOI: 10.1080/00423110701867299. 11

[82] J. Wong. *The Theory of Ground Vehicles*, Wiley, 2001. 11

[83] L. Sun. Simulation of pavement roughness and IRI based on power spectral density. *Mathematics and Computers in Simulation*, 61(2):77–88, 2003. DOI: 10.1016/s0378-4754(01)00386-x. 11

[84] W. Hudson, D. Halbach, J. Zaniewski, et al. Root-mean-square vertical acceleration as a summary roughness statistic. *Measuring Road Roughness and its Effects on User Cost and Comfort*, 1985. DOI: 10.1520/stp34592s. 12

[85] O. Kropáč and P. Múčka. Be careful when using the international roughness index as an indicator of road unevenness. *Journal of Sound and Vibration*, 287(4):989–1003, 2005. DOI: 10.1016/j.jsv.2005.02.015. 12

[86] O. Kropáč and P. Múčka. Estimation of road waviness using the IRI algorithm. *Strojnícky Casopis—Journal of Mechanical Engineering*, 55(5):308–313, 2004. 12, 13

[87] Z. Wu, S. Chen, L. Yang, et al. Model of road roughness in time domain based on rational function. *Transactions of Beijing Institute of Technology*, 29(9):795–798, 2009. 14

[88] B. Heißing and M. Ersoy. *Chassis Handbook: Fundamentals, Driving Dynamics, Components, Mechatronics, Perspectives*, 2013. 15, 20

[89] K. Hong, D. Jeon, W. Yoo, et al. A new model and an optimal pole-placement control of the macpherson suspension system. *SAE Technical Papers*, 1999. DOI: 10.4271/1999-01-1331. 15, 17

[90] H. Pacejka. *Tire and Vehicle Dynamics*, Elsevier, 2005. 18

[91] E. Bakker. *Tyre Modeling for Use in Vehicle Dynamics Studies*, 1987. DOI: 10.4271/870421.

[92] H. Pacejka. The magic formula tyre model. *Proc. of the 1st International Colloquium on Tyre Models for Vehicle Dynamics Analysis*, Delft, Netherlands, 1991. DOI: 10.1080/00423119208969994. 18

[93] T. Gillespie. Fundamentals of vehicle dynamics. *General Motors Institute*, 2000. DOI: 10.4271/r-114. 19

[94] R. Rajamani. *Vehicle Dynamics and Control*, Springer Science, 2006. DOI: 10.1007/978-1-4614-1433-9. 19

[95] H. Li, J. Yu, C. Hilton, et al. Adaptive sliding-mode control for nonlinear active suspension vehicle systems using t–s fuzzy approach. *IEEE Transactions on Industrial Electronics*, 60(8):3328–38, 2013. DOI: 10.1109/tie.2012.2202354. 19

[96] C. Kim, P. Ro, and H. Kim. Effect of the suspension structure on equivalent suspension parameters. *Proc. of the Institution of Mechanical Engineers, Part D: Journal of Automobile Engineering*, 213(5):457–70, 1999. DOI: 10.1243/0954407991527026. 20

[97] W. Aboud, S. M. Haris, and Y. Yaacob. Advances in the control of mechatronic suspension systems. *Journal of Zhejiang University-Science C*, 15(10):848–860, 2014. DOI: 10.1631/jzus.c14a0027. 22

[98] B. Spencer, S. Dyke, M. Sain, et al. Phenomenological model for magnetorheological dampers. *Journal of Engineering Mechanics*, 123(3):230–238, 1997. DOI: 10.1061/(asce)0733-9399(1997)123:3(230). 23

[99] X. Dong, M. Yu, C. Liao, et al. Comparative research on semi-active control strategies for magneto-rheological suspension. *Nonlinear Dynamics*, 59(3):433–453, 2010. DOI: 10.1007/s11071-009-9550-8. 23, 86

[100] X. Song, M. Ahmadian, S. C. Southward. Modeling magnetorheological dampers with application of nonparametric approach. *Journal of Intelligent Material Systems and Structures*, 16(5):421–32, 2005. DOI: 10.1177/1045389x05051071. 23

[101] Y. G. Lei, Z. J. He, Y. Y. Zi, et al. Fault diagnosis of rotating machinery based on multiple anfis combination with gas. *Mechanical Systems and Signal Processing*, 21:2280–94, 2007. DOI: 10.1016/j.ymssp.2006.11.003. 27, 28

[102] Y. Qin, M. Dong, F. Zhao, et al. Road profile classification for vehicle semi-active suspension system based on adaptive neuro-fuzzy inference system. *IEEE Control Decision Conference*, Osaka, Japan, 2015. DOI: 10.1109/cdc.2015.7402428. 28, 43

[103] H. Peng, F. Long, and C. Ding. Feature selection based on mutual information criteria of max-dependency, max-relevance, and min-redundancy. *IEEE Transactions on Pattern Analysis and Machine Intelligence*, 27(8):1226–1238, 2005. DOI: 10.1109/tpami.2005.159. 33

[104] C. Ding and H. Peng. Minimum redundancy feature selection from microarray gene expression data. *Journal of Bioinformatics and Computational Biology*, 3(02):185–205, 2005. DOI: 10.1109/csb.2003.1227396. 33

[105] D. Specht. Probabilistic neural networks. *Neural Networks*, 3(1):109–118, 1990. DOI: 10.1016/0893-6080(90)90049-q. 34

[106] ISO. Mechanical vibration-road surface profiles-reporting of measured data. *International Organization for Standardization*, 1995. 48

[107] L. Breiman. Random forests. *Machine Learning*, 45(1):5–32, 2001. DOI: 10.1023/A:1010933404324. 48

[108] K. Guo, 1998–0148-7191: *SAE Technical Paper*, 1993. 54

[109] Y. Shtessel, M. Taleb, and F. Plestan. A novel adaptive-gain supertwisting sliding mode controller: Methodology and application. *Automatica*, 48(5):759–769, 2012. DOI: 10.1016/j.automatica.2012.02.024. 63, 64, 65

[110] J. Moreno and M. Osorio. Strict Lyapunov functions for the super-twisting algorithm. *IEEE Transactions on Automatic Control*, 57(4):1035–1040, 2012. DOI: 10.1109/tac.2012.2186179. 64

[111] G. Cooper and C. Mcgillem. *Probabilistic Methods of Signal and System Analysis*, Holt, Rinehart and Winston, 1971. 70, 71

[112] Georgec. *Analytical Design of Linear Feedback Controls*, John Wiley & Sons, 1957. 70

[113] J. Zhao, P. Wong, Z. Xie, and X. Ma. Chassis integrated control for active suspension system, active front steering and direct yaw moment using hierarchical strategy. *Vehicle System Dynamics*, 55, 1:72–103, 2017. DOI: 10.1080/00423114.2016.1245424. 73

[114] Y. Huang, A. Khajepour, H. Ding, et al. An energy-saving set-point optimizer with a sliding mode controller for automotive air-conditioning/refrigeration systems. *Applied Energy*, 188:576–585, 2017. DOI: 10.1016/j.apenergy.2016.12.033. 80

[115] K. Deb. *Multi-Objective Optimization using Evolutionary Algorithms*, John Wiley & Sons, Chichester, 2001. DOI: 10.1007/978-0-85729-652-8_1. 80, 81

[116] C. Coello, G. Lamont, and V. Veldhuizen. *Evolutionary Algorithms for Solving Multi-Objective Problems*, Springer, New York, 2007. DOI: 10.1007/978-1-4757-5184-0. 81

[117] J. Schaffer. Multiple objective optimization with vector evaluated genetic algorithms. *Proc. of the 1st International Conference on Genetic Algorithms*, 1985. 82

[118] N. Srinivas and K. Deb. Muiltiobjective optimization using nondominated sorting in genetic algorithms. *Evolutionary Computation*, 2(3):221–248, 1994. DOI: 10.1162/evco.1994.2.3.221. 82

[119] C. Fonseca and P. Fleming. Genetic algorithms for multiobjective optimization: Formulation discussion and generalization. *Proc. of the ICGA*, 1993. 82

[120] J. Horn, N. Nafpliotis, and D. Goldberg. A niched pareto genetic algorithm for multiobjective optimization; proceedings of the evolutionary computation. *IEEE World Congress on Computational Intelligence*, 1994. DOI: 10.1109/icec.1994.350037. 82

[121] K. Deb, A. Pratap, S. Agarwal, et al. A fast and elitist multiobjective genetic algorithm: NSGA-II. *IEEE Transactions on Evolutionary Computation*, 6(2):182–197, 2002. DOI: 10.1109/4235.996017. 82

[122] E. Zitzler and L. Thiele. Multiobjective evolutionary algorithms: A comparative case study and the strength pareto approach. *IEEE Transactions on Evolutionary Computation*, 3(4):257–271, 1999. DOI: 10.1109/4235.797969. 82

[123] D. Corne, J. Knowles, and M. Oates. The pareto envelope-based selection algorithm for multiobjective optimization. *Proc. of the Parallel Problem Solving from Nature PPSN*, 2000. DOI: 10.1007/3-540-45356-3_82. 82

[124] Y. Qin, M. Dong, R. Langari, et al. Adaptive hybrid control of vehicle semiactive suspension based on road profile estimation. *Shock and Vibration*, Article no. 636739, 2015. DOI: 10.1155/2015/636739. 85

[125] M. Ahmadian and N. Vahdati. Transient dynamics of semiactive suspensions with hybrid control. *Journal of Intelligent Material Systems and Structures*, 17(2):145–153, 2006. DOI: 10.1177/1045389x06056458. 86

[126] P. Nugroho, W. Li, H. Du, et al. An adaptive neuro fuzzy hybrid control strategy for a semiactive suspension with magneto rheological damper. *Advances in Mechanical Engineering*, 2014. DOI: 10.1155/2014/487312. 86

[127] E. Camacho and C. Alba. *Model Predictive Control*, Springer, 2013. DOI: 10.1007/978-0-85729-398-5_2. 93, 102

[128] T. Gordon and R. Sharp. On improving the performance of automotive semi-active suspension systems through road preview. *Journal of Sound and Vibration*, 217(1):163–182, 1998. DOI: 10.1006/jsvi.1998.1766. 93, 94

[129] P. Scokaert and J. Rawlings. Constrained linear quadratic regulation. *IEEE Transactions on Automatic Control*, 43(8):1163–1169, 1998. DOI: 10.1109/9.704994. 93

[130] J. Lee and P. Khargonekar. Constrained infinite-horizon linear quadratic regulation discrete-time systems. *IEEE Transactions on Automatic Control*, 52(10):1951–1958, 2007. DOI: 10.1109/tac.2007.906239.

[131] Y. Huang, S. Fard, M. Khazraee, et al. An adaptive model predictive controller for a novel battery-powered anti-idling system of service vehicles. *Energy*, 127:318–327, 2017. DOI: 10.1016/j.energy.2017.03.119.

[132] Y. Huang, A. Khajepour, F. Bagheri, et al. Modelling and optimal energy-saving control of automotive air-conditioning and refrigeration systems. *Proc. of the Institution of Mechanical Engineers, Part D: Journal of Automobile Engineering*, 231(3):291–309, 2017. DOI: 10.1177/0954407016636978. 93

[133] J. Rawlings and D. Mayne. *Model Predictive Control: Theory and Design*, Nob Hill Publishing, 2009. 93

[134] Y. Huang, A. Khajepour, F. Bagheri, et al. Optimal energy-efficient predictive controllers in automotive air-conditioning/refrigeration systems. *Applied Energy*, 184:605–618, 2016. DOI: 10.1016/j.apenergy.2016.09.086. 93

[135] Y. Huang, A. Khajepour, and H. Wang. A predictive power management controller for service vehicle anti-idling systems without a priori information. *Applied Energy*, 182:548–557, 2016. DOI: 10.1016/j.apenergy.2016.08.143. 93

[136] B. Cho. Active suspension controller design using MPC with preview information. *KSME International Journal*, 13(2):168–174, 1999. DOI: 10.1007/bf02943668. 94

[137] R. Mehra, J. Amin, J. Hedrick, et al. Active suspension using preview information and model predictive control. *Proc. of the Control Applications*, 1997. DOI: 10.1109/cca.1997.627769.

[138] S. Gopalasamy, C. Osorio, J. Hedrick, et al. Model predictive control for active suspensions controller design and experimental study. *ASME DSCD*, 1997.

[139] M. Donahue. *Implementation of an Active Suspension, Preview Controller for Improved Ride Comfort*. University of California at Berkeley, 1998. 94

[140] M. Canale, M. Milanese, and C. Novara. Semi-active suspension control using fast model-predictive techniques. *IEEE Transactions on Control Systems Technology*, 14(6):1034–1046, 2006. DOI: 10.1109/tcst.2006.880196. 94

[141] M. Ahmed and F. Svaricek. Preview control of semi-active suspension based on a half-car model using fast fourier transform. *IEEE*, New York, 2013. DOI: 10.1109/ssd.2013.6564120. 94

[142] N. Giorgetti, A. Bemporad, H. Tseng, et al. Hybrid model predictive control application towards optimal semi-active suspension. *IEEE*, New York, 2005. DOI: 10.1109/isie.2005.1528942. 94

[143] N. Giorgetti, A. Bemporad, H. Tseng, et al. Hybrid model predictive control application towards optimal semi-active suspension. *International Journal of Control*, 79(05):521–533, 2006. DOI: 10.1080/00207170600593901. 94

[144] D. Mayne, J. Rawlings, C. Rao, et al. Constrained model predictive control: Stability and optimality. *Automatica*, 36(6):789–814, 2000. DOI: 10.1016/S0005-1098(99)00214-9. 94

[145] L. Wang. *Model Predictive Control System Design and Implementation Using Matlab®*, Springer, 2009. DOI: 10.1007/978-1-84882-331-0. 94

[146] F. Borrelli, A. Bemporad, and M. Morari. *Predictive Control for Linear and Hybrid Systems*, Cambridge University Press, 2017. DOI: 10.1017/9781139061759. 97, 98, 101

[147] W. Heemels, D. Schutter, and A. Bemporad. Equivalence of hybrid dynamical models. *Automatica*, 37(7):1085–1091, 2001. DOI: 10.1016/s0005-1098(01)00059-0. 99, 101

[148] A. Bemporad and M. Morari. Control of systems integrating logic, dynamics, and constraints. *Automatica*, 35(3):407–427, 1999. DOI: 10.1016/s0005-1098(98)00178-2. 101

[149] A. Bemporad. Efficient conversion of mixed logical dynamical systems into an equivalent piecewise affine form. *IEEE Transactions on Automatic Control*, 49(5):832–838, 2004. DOI: 10.1109/tac.2004.828315. 101

[150] D. Mignone. Control and estimation of hybrid systems with mathematical optimization. *Technische Wissenschaften Eth Zürich*, Nr. 14520, 2002. DOI: 10.3929/ethz-a-004279802. 102

[151] T. Ndel, T. Johansen, and A. Bemporad. An algorithm for multi-parametric quadratic programming and explicit MPC solutions. *Automatica*, 39(3):489–497, 2003. DOI: 10.1016/s0005-1098(02)00250-9. 102

[152] A. Hac and I. Youn. Optimal semi-active suspension with preview based on a quarter car model. *Journal of Vibration and Acoustics*, 114(1):84–92, 1992. DOI: 10.23919/acc.1991.4791404. 102

[153] L. Fberg. Yalmip: A toolbox for modeling and optimization in Matlab. *Proc. of the Computer Aided Control Systems Design*, 2004. DOI: 10.1109/cacsd.2004.1393890. 107

[154] F. Zhao, M. Dong, Y. Qin, et al. Adaptive neural networks control for camera stabilization with active suspension system. *Advances in Mechanical Engineering*, 7(8):1687814015599926, 2015. DOI: 10.1177/1687814015599926. 107

[155] F. Zhao, M. Dong, Y. Qin, et al. Adaptive neural-sliding mode control of active suspension system for camera stabilization. *Shock and Vibration*, 2015. DOI: 10.1155/2015/542364. 107

[156] N. Louam, D. Wilson, and R. Sharp. Optimal control of a vehicle suspension incorporating the time delay between front and rear wheel inputs. *Vehicle System Dynamics*, 17(6):317–336, 1988. DOI: 10.1080/00423118808968909. 111

Authors' Biographies

YECHEN QIN

Yechen Qin is currently a Postdoctoral Fellow of mechanical engineering at the Beijing Institute of Technology, where he received his B. Eng and Ph.D. in 2010 and 2016, respectively. From 2013–2014, he studied at Texas A&M University as a visiting Ph.D. student. From 2017–2018, he studied at the University of Waterloo as a visiting scholar. His research interests include vehicle dynamics control, road estimation, and in-wheel motor vibration control.

HONG WANG

Hong Wang is currently a research associate of Mechanical and Mechatronics Engineering at the University of Waterloo. She received her Ph.D. from the Beijing Institute of Technology in China in 2015. Her research focuses on component sizing, modeling of hybrid powertrains, and energy management control strategies design for hybrid electric vehicles; intelligent control theory and application; and autonomous vehicles.

YANJUN HUANG

Yanjun Huang is a Postdoctoral Fellow at the Department of Mechanical and Mechatronics Engineering at the University of Waterloo, where he received his Ph.D. in 2016. His research interest is primarily focused on the vehicle holistic control in terms of safety, energy-saving, and intelligence, including vehicle dynamics and control, HEV/EV optimization and control, motion planning and control of connected and autonomous vehicles, and human-machine cooperative driving. He has published several books and over 50 papers in journals and conferences. He currently serves as an associate editor and editorial board member of *IET Intelligent Transport System*, *SAE International Journal of Commercial Vehicles*, *International Journal of Vehicle Information and Communications*, *Automotive Innovation*, *AIME*, etc.

XIAOLIN TANG

Xiaolin Tang received a B.S. in mechanics engineering and an M.S. in vehicle engineering from Chongqing University, China, in 2006 and 2009, respectively. He received a Ph.D. in mechanical engineering from Shanghai Jiao Tong University, China, in 2015. He is currently an Associate Professor at the State Key Laboratory of Mechanical Transmissions and at the Department of Automotive Engineering, Chongqing University, Chongqing, China. He is also a committeeman of Technical Committee on Vehicle Control and Intelligence of Chinese Association of Automation (CAA). He has led and has been involved in more than 10 research projects, such as National Natural Science Foundation of China, and has published more than 20 papers. His research focuses on Hybrid Electric Vehicles (HEVs), vehicle dynamics, noise and vibration, and transmission control.

Printed in the United States
by Baker & Taylor Publisher Services